Solar Installations

Practical applications for the built environment

LARS ANDRÉN

London • Sterling, VA

First published by James & James (Science Publishers) in 2003

Reprinted by Earthscan in 2007

Copyright © 2003 Lars Andrén

ISBN: 978-1-902916-45-3

Translated by Jill Gertzén
Typeset by Saxon Graphics, Derby
Printed and bound in the UK by CPI Antony Rowe, Eastbourne
Cover photos courtesy of Lars Andrén and (4th from top) Powerlight Corporation

For a full list of publications please contact:
Earthscan
8–12 Camden High Street
London, NW1 0JH, UK
Tel: +44 (0)20 7387 8558
Fax: +44 (0)20 7387 8998
Email: earthinfo@earthscan.co.uk
Web: www.earthscan.co.uk

22883 Quicksilver Drive, Sterling, VA 20166-2012, USA

Earthscan publishes in association with the International Institute for Environment and Development.

A catalogue record for this book is available from the British Library.

Contents

Preface

'Words are like the sun's rays – the more concentrated they are the more they burn.'
Source: Swärd G. and K.

The sun is of essential importance for mankind. Historically it has been not only a vital source of energy but also a symbol for life, religions, myths and much more. However, our modern society finds it difficult to appreciate the sun. This free, abounding energy is an enormous resource that is not fully utilized. I am convinced that solar energy is essential for the achievement of an ecologically sustainable society, wherever we may be on earth. Solar energy will play an important role in our future energy supply, to the advantage of both the environment and the economy.

For most of my working life I have worked, in one way or another, with the use of the sun's radiant energy. To have had the privilege of working with the sun as a 'means of support' has both felt morally right and been unbelievably inspiring. I have had the opportunity of following a technology's birth, introduction and breakthrough onto the heating market. My experience is presented in this book. It should be used as a work of reference, in which various solar actors' knowledge is further disseminated with the aims of increasing the general level of knowledge on the subject, of giving solar energy a chance, and of promoting the worthy expansion that the technology deserves today.

The book focuses on water-based solar heating technology, with information on various areas of use. It begins with a general overview of the characteristics of solar energy. This is followed by more detailed information on water-based solar heating technology and systems, including key ratios and practical examples to help the designer, the installer and the end-user to decide on the applicability of solar heating in different projects. Readers are given a clear picture, with the possibility of making a preliminary study of the technology, system and economy themselves. One of the book's main aims is to give the different participants in the building and ventilation, heating and sanitation trades an opportunity to understand and use solar heating. The book also looks at other solar energy technologies. The reader is given an orientation in passive solar and solar air heating technologies, and there is also an extensive section on solar electricity.

The book fills in the gaps in information within the solar heating area, and can be used as a springboard into the subject.

I hope it makes rewarding reading.

Lars Andrén
www.drivkraft.nu

Note to readers: Every country has its own regulations in heating, ventilation and sanitation. Always find out about the rules and regulations in force in your country, and follow these. In some respects the book takes up typical Swedish conditions. Always check with a local tradesman about what is accepted practice in your country.

For ease of reference, a value in euros has been listed in parentheses after sums in Swedish krona, converted at a rate of 9SKr to €1.

Acknowledgements

A book does not write itself. Thank you, Ulla and my children, for your support and inspiration, and for cheering me on. An extra large thank you to those of you who have helped me in my work: the publishers and their staff, and everybody who has checked the content. Special thanks to: Bengt Perers, Chris Bales and Lars Broman at Högskolan Dalarna; Lars Stolt and Olle Lundberg at the Ångström Solar Centre, Uppsala University; Conny Ryytty at the Swedish Energy Administration; Jan-Olof Dalenbäck at Chalmers University of Technology, Göteborg; Peter Kovacs at the Swedish National Testing and Research Station, Borås; Gunnar Lennermo of Energianalys; Christer Nordström of Nordström Arkitekter AB, Anders Axelsson; and all the rest of you who have been involved and have given invaluable help. Above all I should especially like to thank Jill Gertzén, who translated the manuscript into English, and whose contribution has been invaluable for the completion of the project. I must also mention Guy Robinson of James & James, whose great interest in the book has contributed to the form of the English edition.

■ 1. Introduction

1.1 Solar radiation

The energy radiated by the sun is a result of thermonuclear fusion, in which hydrogen is transformed into helium. This transformation involves a loss of mass, which is converted into energy. Box 1.1 lists some facts about the sun.

Box 1.1 Facts about the sun

The sun is our nearest star, and with its immense size governs the movements of the planets by its force of gravity. The sun is estimated to be 4.6×10^9 years old. Its radius is 696,000 km, and it consists of 71% hydrogen (H), 27% helium (He) and 2% other elements. Its weight can be estimated as 1989×10^{30} metric tonnes (333,000 times as heavy as the earth).

The energy-producing process in the sun is the transformation of hydrogen atoms into helium atoms. The transformation involves a loss of mass, which is converted into large amounts of energy emitted in the form of electromagnetic radiation. The total radiation from the sun is 3.8×10^{26} W. Of this amount about 1.7×10^{17} W reaches the earth, and this is (1990) 15,000 times the energy that is consumed on earth.

Solar irradiance outside the earth's atmosphere has an average power of 1370 W/m^2, measured on a surface perpendicular to the direction of the radiation when the earth is at its mean distance from the sun. This value is called the *solar constant*. Only part of the sun's energy reaches the surface of the earth. Some is reflected back, and some is absorbed by water vapour, ozone and carbon dioxide in the atmosphere (Figure 1.1). The molecules in the atmosphere scatter some of the sun's rays, dimming direct sunlight but creating the radiant blue sky. The maximum solar radiation that reaches the surface of the earth is about 1000 W/m^2, including both direct and diffuse radiation.

The variation in global radiation is large, as can be seen by comparing the Mediterranean area, the pure desert areas, and the northern parts of Scandinavia (Kiruna in Sweden) (Table 1.1). The area around the equator receives more energy than northerly or southerly latitudes, because the angle of incidence is higher and the distance through the atmosphere is shorter. There is also less cloud at the equator than in Northern Europe, for example. In the desert areas of the earth the average figure for solar insolation is 2200 kW/m^2 per year, whereas Sweden has an average solar insolation of slightly over 1000 kW/m^2 per year.

Table 1.1 Global radiation

Place	kWh/m^2 per year
Kiruna	870
Stockholm	980
Malmö	1000
Paris	1000
Mediterranean area	1400–1800
Sahara/Arizona	2300–3400

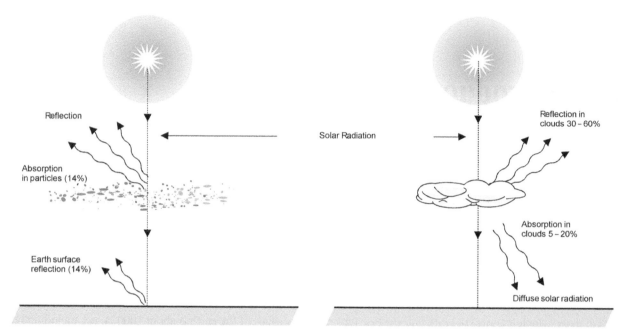

FIGURE 1.1 *How the atmosphere affects solar radiation*

Solar radiation that reaches the surface of the earth contains wavelengths from ultraviolet to infrared, but a significant part of the energy is in the infrared range. For good utilization of solar radiation it is necessary for all wavelengths to be able to be absorbed. Materials and surfaces with high absorptance and low emittance are preferable.

1.2 Available energy

The amount of energy that can be utilized is affected by supply, location, reduction in the atmosphere, reflection and absorption in the clouds, and the angle that the object used to absorb the energy makes with the horizontal. Solar radiation is normally classed as direct (beam) or diffuse. *Direct sunlight* causes strong shadows; indirect or *diffuse sunlight* is reflected from clouds or other objects. The sum of direct and diffuse solar radiation is the radiation we can make use of. It is called *global radiation*. The amount of radiation that

reaches the earth's surface is influenced largely by the local weather. The direct radiation can be reflected and absorbed in the clouds, so that on occasions it is practically non-existent.

For example, in the south of Sweden (areas near the coast, around latitude 56° N) global radiation is slightly over 1000 kW/m^2 per year, whereas the corresponding figure for the northern part of the country (around the Arctic Circle) is slightly over 800 kW/m^2 per year (on a horizontal plane). The total radiation increases by about 25% if the absorbing object is tilted between 30° and 45° from the horizontal plane towards the south. In Sweden, the amount of solar energy on a 30° south-sloping surface varies from the best positions along the coasts (1250 kW/m^2 per year) to the worst inland conditions in the northern part of the country (900 kW/m^2 per year). See Figure 1.2. Table 1.2 shows all the incident (direct + diffuse and ground-reflected) solar radiation (per m^2) on a south-oriented surface with 30° tilt to the horizontal plane, given in kWh/day.

TABLE 1.2 Solar radiation for places on or near latitude 60° N (Sweden)

Month	Clear days (kWh/day)	Scattered clouds (kWh/day)	Overcast days (kWh/day)
January	1.42	0.92	0.28
February	3.28	2.24	0.76
March	5.28	3.70	1.48
April	7.02	5.34	2.30
May	8.08	6.36	2.94
June	8.52	6.84	3.26
July	8.34	6.66	3.12
August	7.54	5.84	2.60
September	6.06	4.50	1.82
October	4.14	2.92	1.06
November	2.10	1.38	0.44
December	0.98	0.62	0.18

Source: Gunnar Lennermo, Energianalys, Sweden

Of the radiation that reaches Sweden (approximately 900–1250 kWh/m² annually), a solar collector of slightly under 50% efficiency can produce approximately 400 kWh/m² annually (an average value for a standard flat-plate collector of good quality). With slightly over 2,500,000 m² of solar collectors, 1 TWh of heat can be produced!

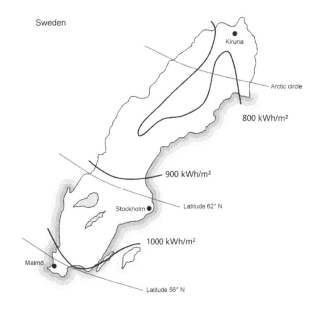

FIGURE 1.2 *Variations of global radiation on a horizontal surface in Sweden. There are large variations between the seasons, but the measurements for the same month can differ from one year to another. In Sweden there is generally more direct radiation (clear skies) along the coasts and in the area around Lake Vänern Source: Svenska Solenergiföreningen (1991)*

■ 2. The solar heating system

Water-based solar heating is the dominant application of solar energy in Europe. Ever since the oil crises of the 1970s there has been intensive research in the field, and as a result a wide range of technologies is available today. There are both more and less well-established solar heating systems for different areas of use and target groups – everything from small domestic hot water (DHW) systems to large-scale solar heating plants for use in district heating.

2.1 Orientation and output

The incident angle of the sun's rays on a flat stationary solar collector varies with time. This variation affects the amount of solar radiation that reaches the solar collector. Usable radiation decreases when the orientation deviates from the south, and is also affected by the tilt (Figure 2.1).

An important detail in design work is to find suitable places for solar collectors that give the required heat output and are acceptable from the point of view of cost. A basic question is: at what angle and orientation should the solar collectors be placed to utilize as much as possible of the solar radiation?

The optimal position for a solar collector is facing south (in the northern hemisphere) and is dependent on several factors such as the latitude, the load and whether the collector is 'over-dimensioned' for summer conditions. For a constant load that is met by the collector without significant overproduction during summer, the optimal tilt angle is 10–15° less than the latitude. If the load is greater in winter than in summer, as in solar combisystems, then the optimum angle is greater. Table 2.1 shows that the optimal placing of solar collectors in Sweden (based on glazed flat-plate solar collectors) is facing directly south with a 45° tilt, but that a south-east or south-west orientation does not appreciably affect the output. However, it is important that the solar collectors are not orientated directly to the east or west. If the deviation from south is not more than to the south-west or south-east it is advantageous for the solar collectors to have a smaller tilt. Vertical solar collectors give a larger output than horizontally placed ones during the spring and the autumn. With the help of the table, correction can be made to calculate the loss due to a given divergence from optimal placing of the solar collectors.

The solar collector's tilt can be reduced if heat production is intended mainly for summer use.

TABLE 2.1 Relative output from solar collectors with different tilts and orientation (latitude 55°N, Sweden)

Orientation	Tilt			
	15°	30°	45°	60°
S	0.91	0.99	1.00	0.96
SW, SE	0.87	0.92	0.93	0.89
W, E	0.79	0.78	0.75	0.69

Source: Jan-Olof Dalenbäck, Chalmers University of Technology, Göteborg, Sweden

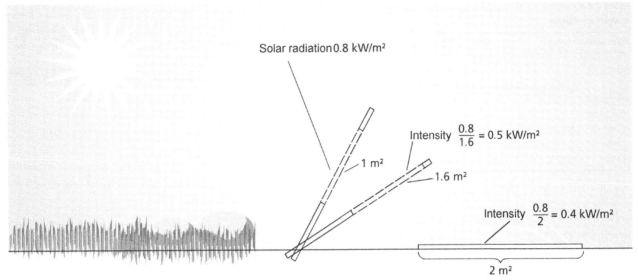

FIGURE 2.1 *The intensity of solar radiation on a surface varies with the angle of the surface in relation to the solar radiation*
Source: Gunnar Lennermo, Energianalys, Alingsås, Sweden

One typical example is unglazed solar collectors for heating outdoor pools.

It is important to avoid all forms of shading. Make a note of the risk of overshadowing from possible dormer windows, or if there is a risk that nearby vegetation might grow and overshadow the solar collector.

2.2 Sizing

The sizing of a solar heating plant is crucial for the yield. The efficiency of the plant and the financial return are determined largely during the design and sizing phase. It is important to decide which demands should be fulfilled, and what determines the size. For optimal output (both technically and economically) the heat load to be covered by the solar heat must be identified, to avoid oversizing (which is not so unusual) and thereby producing fewer kWh for the money invested. An oversized plant gives excess heat during the summer and only a small increase in the coverage of the total heat load. See Figure 2.2.

Table 2.2 lists the normal sizes of solar heating systems in colder climates in Northern Europe.

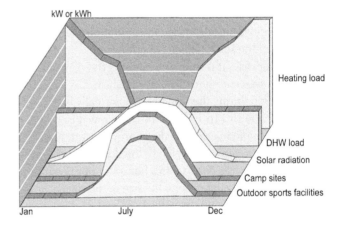

FIGURE 2.2 *Solar radiation during the year is given and should be compared with the heat load to be covered by the solar collectors. In some cases the demand and supply coincide naturally; in others more careful design and sizing are required*

In addition to the heat load to be covered there are other considerations at the design stage. The choice of position and the space for the solar collectors and storage tank can affect the solar collector area. In some cases the desire to replace heat produced at poor efficiency can motivate a somewhat larger solar

TABLE 2.2 Normal sizing of solar heating systems in colder climates in Northern Europe

Solar collector area	DHW system	Combisystem
per person	1–2 m²	2–3 m²
per flat in multi-family dwelling	3–4 m²	5–8 m²
per single-family house	5–8 m²	10–12 m²
Heat store		
per m² solar collector	50–75 litres	75–100 litres
per flat in multi-family dwelling	200–300 litres	300–500 litres
per 100 m² habitable area	300–500 litres	600–800 litres

collector area. Solar heat, together with auxiliary heating as a heating guarantee, can improve the efficiency of the whole system. It is always of interest to combine solar heating with heating sources that have worse efficiency during the low-load period (the six summer months). It is an advantage if solar heating can replace the heat from a boiler during the summer, when the boiler efficiency is lowest. In this way the efficiency of the system is improved, and the operation and economy of the solar heating are optimized.

In other cases it is simpler to identify the demand. For camp sites and sports facilities the heat consumption during the summer can normally be determined fairly precisely. In the same way it is relatively simple to calculate the energy consumption for an outdoor pool. In dwellings, the domestic hot water consumption (60–70 litres hot water per person per day) normally determines the size of the solar collectors. In this connection the heat losses, which can be a large part of the required heat, should not be underestimated. The temperature of the cold water supply also varies greatly during the year.

When considering single-family houses, however, there are other criteria that can affect the solar collector area, for example keeping a basement free from damp during the summer, or being able to turn off a boiler with poor summer efficiency. A solar heating plant can have an important role as a supplement to a heating system. It is normal in multi-family dwellings for the heating control equipment to turn the heat off when the outdoor temperature exceeds +18° C and put it on

when the outdoor temperature is less than +16° C, with a time delay of 1–3 hours. This means that the heating distribution system runs during many nights during the early summer and autumn. Solar heating can be used for heating in this case and thus increase the coverage and the profitability. Heat losses in the DHW circulation system should also be taken into account where applicable. (These can be estimated as 20% of the DHW demand yearly.)

There are a number of computer programs available for sizing solar heating systems. The most common are described briefly in Box 2.1.

2.3 The solar collector

The most important component in a solar heating system is, of course, the solar collector. Over the years, extensive development work has been carried out to find efficient solar collector constructions. Performance has more than doubled and the price has halved over a period of somewhat less than 10 years. The solar collectors of today have an average efficiency of around 50% at 'normal' working temperatures, and this must be regarded as very good.

Solar heating technology is well tried, and there are authorized testing stations in many countries. A state grant or investment support is often coupled to a requirement that the solar collector be tested and approved by an authorized testing station, which guarantees the quality of the technology.

With increased interest and expanding sales, solar collector production will be rationalized and move towards industrial mass production. The materials and the construction will be optimized to give lower production costs. Alternatives will also be available for flexible applications, for example for different roofing materials. The construction of the solar collectors will allow simple and cheap installation.

The development of solar collectors has been intensive ever since the oil crises of the 1970s. Many different types of construction have been presented, tested and developed, and others have been rejected.

Box 2.1 Computer programs: solar energy

Watsun (version 1.2)
Produced by Watsun Simulation Laboratory, University of Waterloo.

Average hourly values are required as climate data, but synthetic hourly values can be created if only monthly values are available. The program has standard models for indoor and outdoor pools with and without active solar heating. There are also options for sizing such things as traditional solar-heated DHW systems and process heat systems.

There is also a version for solar cell systems: WATSUN-PV.

Minsun (version 4)
Developed by IEA.SH (International Energy Agency, Solar Heating), VUAB (Vattenfall Utvecklings AB), and SP after an original version by Studsvik AB, Sweden.

Older programs in DOS. Consists of a solar collector model (UMSORT) and a system model (MINSUN). UMSORT calculates the heat output of a solar collector at a number of working temperatures. MINSUN calculates the operation of a solar heating plant with a seasonal store.

The solar collector models in the program take into account performance, shading in the field, and climate data (average hourly values). The user can choose between flat-plate, concentrated or evacuated solar collectors.

A special version of UMSORT is used by SP for the energy specification of solar collectors. VUAB has developed a Windows version of UMSORT in TRNSYS, which is called WINSUN.

Poolsol
Developed by Teknologiskt Institut, Denmark.

A user-friendly program specially written for outdoor pools with active solar heating and without DHW production. Has detailed models for pools, solar collectors and component parts such as pipe systems. Great freedom of choice for data input means that many alternatives can be studied. The climate data are limited to Denmark.

FCHART
Developed by FCHART Software, Wisconsin, USA.

Good program for estimates. It is possible to calculate many different types of solar heating system. The latest version also contains calculations for pools with and without active solar heating.

For more information: www.fchart.com

Zwembad
Produced by TPD, Delft, Holland.

A detailed simulation program for supercomputers, primarily intended for calculations of pool facilities.

T*sol
A German program for a number of standard types of system for solar heating. Modern, with graphic presentation of systems and results. Climate data are available for a large number of places.

There is also a solar electricity program PV*SOL. For more information: www.valentin.de

TRNSYS
Modular simulation program developed in the USA and Germany. Is frequently used in Europe, and is common in research circles. Applicable to both solar cell and solar heating systems. Detailed system calculations (component models) can be made, but ask a great deal of the user. Building models can be created with several zones. An additional tool, PRESIM, from Sweden makes system construction easier, and the TRANSED tool allows the less familiar user to utilize TRNSYS models in a similar way to simpler programs.

For more information: http://sel.me.wisc.edu/

CoDePro
A program that is particularly interesting for students.

For more information: http://sel.me.wisc.edu

Solar cells
WATSUN-PV
PVSYST
TRNSYS

Others
Many manufacturers and suppliers have their own programs that are used for sizing. As well as those given above, there are a large number of programs available in other countries, such as Denmark (contact secd@eknologisk.dk) and Austria.

The result of this development of the technology can now be seen, in that the flat-plate solar collector completely dominates the market. In order to give insight into the development of the technology and the various constructions, some different types of solar collector are described here.

2.3.1 FLAT-PLATE SOLAR COLLECTORS
During the latter part of the 1980s and during the 1990s highly efficient flat-plate solar collectors were rapidly developed (Figure 2.3). Significant changes in their construction have been the development of absorbers (see Figures 2.4–2.8 and Box 2.2), better-insulated solar

collectors, the development of convection barriers, and the manufacture of solar collectors in larger modules. The production cost per m² for solar collectors that are constructed in large, continuous modular units is generally lower and the efficiency is normally improved. One of the contributing factors to the improved efficiency is that the total heat loss decreases as a result of lower *edge losses*. These are losses of heat through the frame of the solar collector between the cover glass and the insulation. The length of the solar collector frame per m² of area is reduced in large continuous modular units.

Factory-assembled solar collectors are used for both single-family and multi-family dwellings. They can be placed on the ground or be designed for mounting directly on the roof. Collectors for multi-family dwellings are normally manufactured in larger units (up to 15 m²/unit). They are therefore cheaper per m², have better performance, and are easier to install.

During the middle of the 1980s an interesting roof-integrated solar collector was developed in Sweden. This type is assembled on the building site. The bottom layer of the solar collector is placed on the roof construction and the underside insulation is placed on top of it. A vapour barrier (fibreglass sheet or aluminium foil) is laid over the insulation, and the absorber is placed on this. Tempered glass is usually

Box 2.2 Absorber construction

At the beginning of the 1980s, Gränges Aluminium developed an absorber to stand the great stresses (including temperature variations, solar insolation and humidity) that occur in a solar collector. It is constructed of a copper pipe, which is rolled between two sheets of aluminium. In this way good contact is obtained between the surfaces of the absorber fin, the surface receiving the solar radiation, and the pipe in which the heat transfer fluid is transported. To create good heat transfer between the fin and the heat transfer fluid the pipe is rhomboidal, which increases the opportunity for turbulent flow in the pipe. This increases the heat transfer from the pipe to the fluid, thereby decreasing the temperature difference between fin and fluid and improving the efficiency. Rounded ends to the pipes make soldering easier. The absorber surface is anodized for maximum efficiency, which means that high absorptance (0.9–0.98) and low emittance (0.08–0.15) are achieved. Modern anodizing is carried out by a sputtering process.

Another type of absorber is made wholly of copper, in which the fin and the pipe are welded together. The pipe loop is in one piece without soldered joints, which minimizes the risk of leakage.

Box 2.3

A solar collector can work both as a roofing material and as a producer of heat, and thus replace the roof construction. Solar collectors can also replace the wall cladding. Thus both the character and identity of the building may be changed.

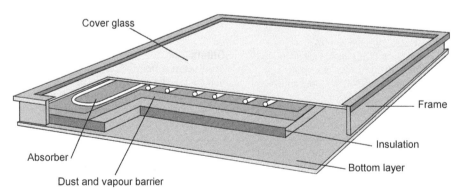

FIGURE 2.3 *A flat-plate solar collector is constructed as follows. A structural frame of aluminium or some sort of metal sheet supports the solar collectors, for example when they are free-standing. A bottom layer is placed in the frame and insulation is laid on top. This in turn is covered by a dust and vapour barrier (such as aluminium foil or fibreglass sheet). The absorber, the most important component in the solar collector, is placed on top of this. The cover material is normally tempered glass but can also be some form of plastic material, such as UV-resistant acrylic plastic of suitable quality. Some solar collector types can have a convection barrier between the absorber and the cover glass*

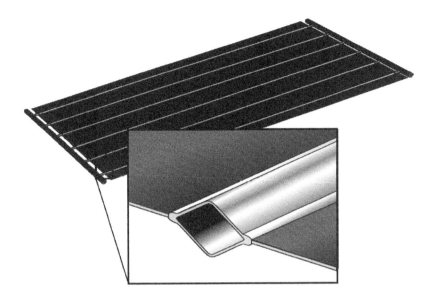

FIGURE 2.4 *Absorber technique in which a copper tube is placed between two sheets of aluminium. A high roll pressure gives a metallurgical contact between the metals. The selective surface coating is sputtered onto the absorber. In sputtering, metal atoms are placed as a surface coating on the absorber with the help of a strong negative charge. The process is carried out in a vacuum chamber with an inert gas. Note that the absorbers are parallel-coupled internally, which is necessary for drainback collectors*

used as the cover material, but different types of plastics (polycarbonate or acrylic) are also found.

The site-built solar collector method has found a natural market for multi-family dwellings. Since the beginning of the 1980s, EKSTA Bostads AB in Kungsbacka has constructed about 6000 m² of roof-integrated solar collectors. Today, this type of site-built solar collector construction can be found even in do-it-yourself circles, where the main target group is single-family home owners. Larger solar heating projects for multi-family dwellings will probably change over to a prefabricated solar collector construction.

In recent years a Swedish company manufacturing prefabricated houses, together with Chalmers University of Technology (Department of Building Services Engineering) and one of Sweden's leading solar heating companies, has developed a new type of solar collector construction that is being marketed in Europe.

The solar collector is of interest because it is manufactured as a finished roof element (Figure 2.9). It is delivered as a complete solar collector element to building sites and is mounted directly on the roof trusses with a building crane. Installation and operation are simple and reliable, and the price is favourable. This type of solar collector has also given good results in test at SP, the Swedish National Testing and Research Station in Borås.

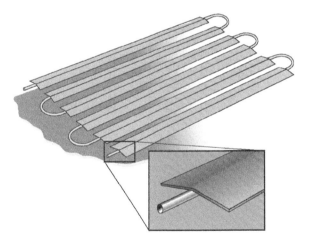

FIGURE 2.5 *All-copper absorbers internally coupled in series. The absorber fin (copper sheet) is ultrasonically welded to a seamless drawn copper tube. The selective surface of the copper absorber is of black chrome, which is applied by an electrolytic process*

Prefabricated solar collectors on the building construction site open up new opportunities for solar heating for both the single-family and multi-family housing markets. The price will decrease, reliability of operation will increase, and installation will be easier.

FIGURE 2.6 *A new type of absorber module has become increasingly popular on the European market. The absorber is made of a pipe bent throughout its length and a finished absorber sheet, delivered by a supplier. The pipe is welded to the metal sheet by a laser welding machine; no extra material is needed. The materials can be copper/copper or aluminium/copper. The absorber sheet can be delivered up to 1250 mm wide and obtained in desired lengths*

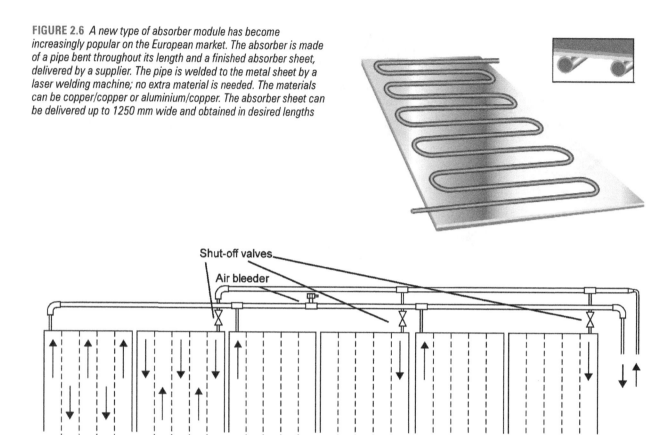

FIGURE 2.7 *Asymmetrical absorber coupling. This is the most common way of coupling solar collector modules, partly because the pipe runs are shorter and installation is simpler. As shown, the supply pipe to the solar collector module on the left is longer, which means that there is a lower flow in this solar collector than in those earlier in the loop. To reduce the difference in flow, the connecting pipe is oversized: in this case (a single-family house system) the pipe is increased from 15 mm to 22 mm at the solar collectors. In many cases a sufficiently good flow can be achieved in the solar collectors by adjusting the pressure in the loop*

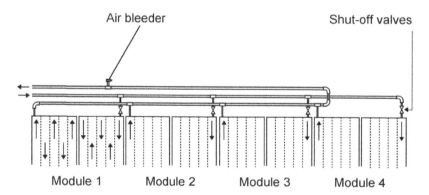

FIGURE 2.8 *Symmetrical absorber coupling. The supply pipes to and return pipes from the solar collector modules are the same length (total lengths), which means that every separate module has the same flow. The connecting pipe to the solar collectors can have the same dimension (with a smaller diameter than with asymmetrical coupling) as the main pipe from the heat store. Note that the air bleeders in the systems are placed differently depending on which method of coupling is chosen*

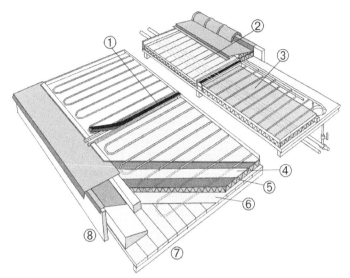

FIGURE 2.9 *Exploded drawing of a solar collector unit. 1, sealing strip of EPDM rubber; 2, tempered cover glass; 3, absorber; 4, aluminium foil; 5, high-temperature insulation; 6, roofing felt; 7, tongue-and-groove boarding; 8, wooden frame*
Source: Derome Träteknik AB, Sweden

2.3.2 DRAINBACK SOLAR COLLECTORS

In a drainback system (Figure 2.10), special demands are made on pipe runs and solar collectors, as well as on the system design. Drainback solar collectors must have relatively large diameter (18–22mm) connecting pipes on the top and bottom edges. When the system is not running, the heat transfer fluid must be completely emptied from the solar collectors and from the part of the main pipe that may be subjected to temperatures below freezing. To ensure drainage the absorbers (pipes) in the solar collectors must be internally parallel-coupled (see Figure 2.4), and the main pipe must slope throughout its length. Most of the solar heating systems sold on the European market are pressurized. The drainback system technology is relatively unusual.

2.3.3 EVACUATED SOLAR COLLECTORS

Evacuated solar collectors have good efficiency, even at high working temperatures. The efficiency at high temperatures increases the area of use to high-temperature systems, and the better efficiency means that a smaller solar collector area is needed for a given heat demand.

The construction of evacuated solar collectors has been developed from fluorescent tube technology (Figure 2.11). The technology has advanced considerably in recent years. This type of solar collector has been installed in both large and small systems, and there is good knowledge of the technology.

The technology is based on an absorber placed in an evacuated glass tube. The vacuum has an insulating function. There are two main operating principles for evacuated solar collectors: dry heat transfer and wet heat transfer. The most common type is the heat pipe, which works with an evaporating and condensing technique in which the heat transfer to the solar circuit is dry: that is, the solar collector has a closed circuit in which the heat is collected in a metal block at one end. From the heat block, heat is transferred to the solar loop.

The construction of a vacuum solar collector with wet heat transfer is similar to that of a flat plate collector. Heat transfer is through an absorber that is in direct contact with the solar loop. Several types of evacuated solar collector contain absorbers and heat transfer fluids similar to those of flat-plate collectors. The complete system, with control equipment, pump,

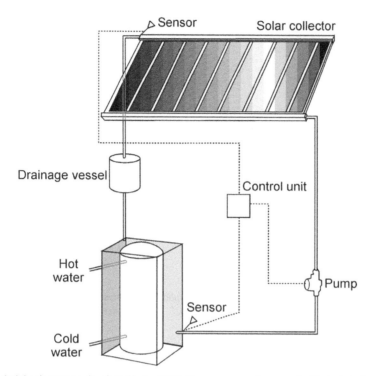

FIGURE 2.10 *In the drainback systems developed in the 1970s it was common to use a double-jacketed tank. The heat transfer fluid is in the outer circuit and fresh water is stored in the inner (normally a DHW tank). When the difference in temperature between the solar collector and the heat store has evened out, the pump in the solar loop is turned off and the solar collectors are drained. The heat transfer fluid is collected in a separate drainage vessel, which must be stored in a frost-free place. It is important that the solar collectors and the parts of the collector pipe that can be exposed to frost have a common lowest point to make sure that they can be emptied*

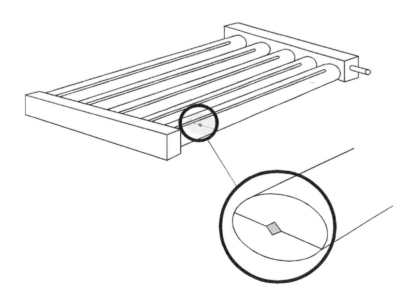

FIGURE 2.11 *Evacuated solar collectors are constructed of separate glass tubes, which are assembled in a frame to the desired size. Generally the heat in the evacuated solar collector (which in itself is a closed circuit) is transferred to the heat transfer fluid in the solar loop*

and piping to and from the solar collectors, is not much different from other solar heating systems.

New types of construction of low-vacuum solar collectors are being developed. In Sweden there are ongoing laboratory experiments using a vacuum as the convection barrier in a flat-plate solar collector. These have double, anti-reflecting coated glass (with spacers). However, evacuated solar collectors still have a relatively small share of the market in Europe, although there has been a notable increase in certain European countries such as Germany. Costs must be forced down for evacuated collectors be able to compete with flat-plate collectors.

2.3.4 CONCENTRATING SOLAR COLLECTORS

See Figures 2.12 and 2.13.

A concentrating solar collector type was used in one of the first large-scale solar heating projects in Sweden (Ingelstad outside Växjö; see Figure 2.14). Solar radiation was concentrated with the help of concave mirrors onto an absorber pipe that was placed at the focus. The experience from the Ingelstad project showed, among other things, that linear-focus (concentrating) solar collectors have limited prospects in Northern Europe. This is principally because the amount of direct solar radiation is far too small in this kind of climate. A conclusion that can be drawn from this project is that linear-focus concentrating solar collectors require advanced technology (where the solar collector follows the path of the sun during the day), which in turn gives unreasonably high investment costs.

Other projects with concentrating solar collectors have been built in places with better conditions, for example in California in the USA. One of the advantages of concentrating collectors is that really high working temperatures are possible. There is a large-scale project in the Mojave desert (east of Los Angeles in the USA) where large concave mirrors heat a heat transfer fluid to 400°C, and this is later used to drive a turbine and generate electricity. In the middle of the 1990s a plant was installed with a power of 675 MW (which can be compared with a small atomic power

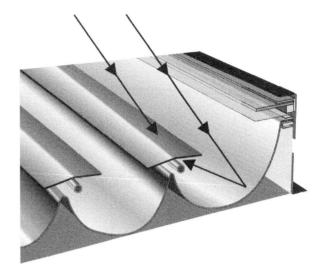

FIGURE 2.12 *In recent years new types of reflecting solar collector have been developed. By using reflectors, both sides of the absorber can be utilized, which reduces the amount of material in the solar collector. The example shown also gives the opportunity for rational/industrial volume manufacture, which should lead to lower construction costs in the long run. There is another type of solar collector with reflectors that is intended for larger projects. A long concave reflector runs along the length of a box, with an absorber placed at the focus of the reflector. This MaReCo solar collector is intended primarily for larger systems such as multi-family dwellings and district heating projects*

plant). The heat provides enough electricity for about 800,000 people. The large-scale projects that have been built in the USA have shown costs of around $0.08 per kWh.

2.3.5 LOW-TEMPERATURE (POOL) SOLAR COLLECTORS

Low-temperature solar collectors have a limited area of use. They can be suitable for heating outdoor pools, as the temperature demand is not so high (15–30°C). The construction of the solar collectors varies, but they are all fairly simple (Figure 2.15). Figure 2.16 shows a completely new type of low-temperature absorber.

Low-temperature solar collectors are normally unglazed and uninsulated, and are thus adapted to low working temperatures. Where solar heat is used for heating pools the low-temperature solar collector is

FIGURE 2.13 *An interesting use for simple concentrating solar collectors is for cooking in areas with a good supply of direct solar radiation, as for example in developing countries. The solar collector requires clear sunshine for the temperatures required for food preparation*

FIGURE 2.14 *In 1978 a solar heating plant with 1325 m² of parabolic solar collectors connected to a seasonal store was built in Ingelstad outside Växjö, Sweden. The solar collectors concentrated the solar radiation on a tube at the focus. The solar heating plant was designed to cover 50% of the heating load in 52 single-family houses nearby*
Photo: Tommy Svensson, Sweden

normally constructed in a material that can resist chlorinated water, as the pool water is used as the heat transfer fluid in the solar circuit. For example, pool solar collectors can be manufactured of EPDM rubber or some sort of UV-resistant polyolefin material. Plastic solar collectors are very sensitive to frost damage, and, if they are exposed to frost risk, the drainback function must be safeguarded. It is important to follow carefully the instructions given by the respective supplier, and the relevant rules and regulations of the country in which the collector is installed.

2.3.6 POSITIONING

Most flat-plate solar collectors can be adapted to different roofing materials, and the structures are also self-supporting. Some manufacturers have large-module solar collectors, which are installed with the help of a mobile crane. Solar collectors for integration in the roofing material normally have complete mounting kits to fix to the battens or boarding. It is important that the level of the solar collectors be adjusted to the roofing material to give natural drainage.

FIGURE 2.15 *A low-temperature solar collector needs neither box, insulation nor cover glass, which gives a simpler construction. Normally the solar collector consists of parallel channels (made of UV-stabilized EPDM rubber or polyolefin plastic) with a common collector tube*

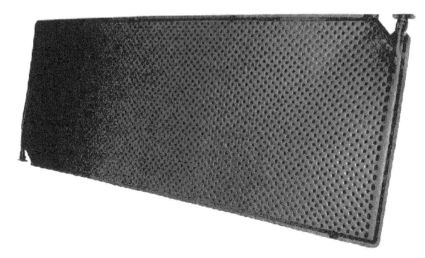

FIGURE 2.16 *A completely new type of absorber was introduced at the end of the 1990s. It is made of textile (woven fibre) with a polymer coating. According to the manufacturer it has good performance and temperature durability. At the time of writing the product is being tested and evaluated by SP (the Swedish National Testing and Research Station) among others*

The supplier's instructions must be followed carefully regarding supervision, operation and safety precautions. When choosing the location of the collector it is important to pay attention to safety, the risk of accidents and the risk of freezing. Fixing, connections and pipe positions should both fulfil the functional and safety requirements and pay attention to aesthetical considerations.

Figure 2.17 shows three common ways of positioning solar collectors on single-family houses.

2.3.7 EFFICIENCY

The efficiency describes the relationship between utilized and supplied energy in a system or in a process of conversion. All types of process have losses that determine the efficiency for a certain energy conversion. The efficiency varies for different processes: that of electric energy can be very high (up to 95%), whereas in other types of heat process it can be limited to 40–50%.

The efficiency of a solar collector compares the incident solar radiation with the energy that the solar collector produces. The energy that is converted from the solar radiation in a solar collector can be expressed in kWh/m^2 based on a given value for the solar radiation (Figure 2.18).

In fact it is more instructive to talk about the system's fuel reduction rather than the solar collector's heat production. By 'removing' the poor efficiency, the energy used for heating can be reduced by considerably more than the solar collector's calculated production. There is a enormous difference if, for example, it is

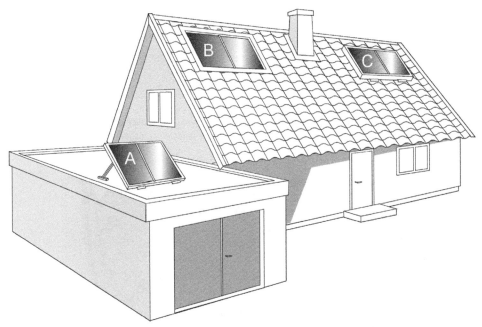

FIGURE 2.17 *Three common ways of positioning solar collectors on single-family houses. (A) Freestanding mounting on a garage roof or similar position makes great demands on the fixing, which must stand wind loads and other stresses. Perforation of the roof surface requires careful waterproofing, especially for flat felt roofs. (B) Solar collectors that are set into the roof tiles have advantages, for example in new buildings, when the solar collector replaces the roof covering material. A metal flashing is placed between the solar collector and the tiled roof. There is an aesthetical advantage, but also a technical disadvantage as the pipe connections and sensor unit are difficult to get at. (C) Solar collectors placed on top of the roof tiles or other roofing materials are often a cost-effective solution. Most solar collector manufacturers have complete mounting kits for this purpose. The pipe connections and the sensor units are easily accessible. Installation near the ridge means a smaller snow load, and the risk of overshadowing by growing vegetation is also reduced. A space between the solar collector and the roofing material is advantageous to allow melting snow, leaves etc. to pass underneath. It makes no difference to the operation and performance whether a good, well-insulated solar collector is placed recessed in the roof or freestanding*

calculated that the solar collector produces 400 kWh/m²
when the real saving in fuel can amount to 800 kWh/m²
per year. This is perfectly possible in systems in which
the solar heating replaces a boiler with a 50% efficiency
in summer.

To obtain comparative values for different makes of
collector, the Swedish solar heating industry has agreed
on a rule of thumb based on an equation that
determines the energy output theoretically under the
same conditions. This is described in Box 2.4.

The performance and quality of solar collectors that
have been tested by an accredited research or testing
institute are well documented. Their tests define the
efficiency of the collector; the efficiency equation is
given in Box 2.5. Box 2.6 lists some other European
testing stations.

2.3.8 HEAT PRODUCTION

The annual heat production from different types of
solar collector can be determined theoretically provided
all the input data (such as orientation and slope, solar
insolation for the location, and relevant efficiency) are
available. Figure 2.19 shows the energy flow and heat

losses in a solar collector, and Figure 2.20 shows how
the heat production from a collector varies with the
amount of solar radiation and the size of the heat losses.

A modern flat-plate solar collector produces
approximately 400 kWh/m² annually. By reducing the
working temperature in the solar collector and thereby
the difference in temperature between the collector
and the ambient air, the efficiency and heat production
can be increased. The working temperature in the solar
loop can therefore be just as decisive for the solar
collector's efficiency as the collector's actual
performance. Using Karlsson's equation (Box 2.4), the
theoretical heat production can be calculated and the
variables changed as desired. Although solar insolation
(expressed in W/m²) varies throughout the country,
most of the calculations are based on an insolation of
between 700 and 800 W/m². The operating time is
calculated as about 1000 hours per year, which gives an
annual supply of 700–800 kWh/m² in Sweden (56°N).
With 50% efficiency (which is normal with a
temperature difference of 50°C) the annual heat output
from a normal solar collector will therefore be
approximately 350–450 kWh/m². Highly efficient

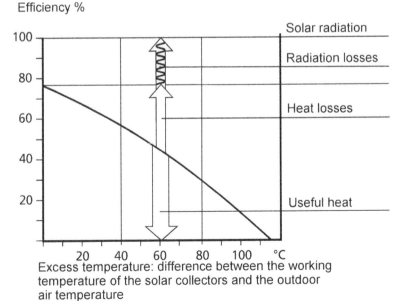

Efficiency %

FIGURE 2.18 *The efficiency graph shows how
much useful heat a solar collector produces under
given conditions. The output for a given amount of
solar radiation is expressed as a percentage and is
the result of the radiation and heat losses that occur
with different disparities between the working
temperature of the solar collector (average of the
supply and return) and the ambient temperature*

Box 2.4 Energy and heat calculation

An equation worked out by Björn Karlsson at Vattenfall Utveckling AB, Älvkarleby, Sweden, can be used to calculate the heat production from a solar collector manually:

annual heat production = absorbed solar radiation – heat losses

or, expressed as an equation:

$$E = (T \times A \times 705) - (U \times k)$$

where E is the annual heat output from the solar collectors; T is the transmittance or translucence of the cover sheet; A is the absorptance of the absorber; U is the effective loss coefficient of the solar collectors; and k is a temperature constant, a unit that relates the number of degree hours per year to the solar collector's average working temperature. The figure of 705 kWh/m²/year is a standard value, which has been derived from data measured in Stockholm, Sweden (lat. 60° N), in 1986. (See also k_e in efficiency equation in Box 2.5.)

For example:

$$E = (T \times A \times 705) - (U \times 37)$$

The solar collector's average temperature is 40°.

$$E = (T \times A \times 705) - (U \times 46)$$

The solar collector's average temperature is 50°.

The values of T, A and U are obtained from tests on the solar collector. Typical values for a normal single-family house solar collector are $T = 0.85$, $A = 0.90$ and $U = 4.5$ (W/m², °C).

$$E(40°) = (0.85 \times 0.90 \times 705) - (4.5 \times 37) = 373 \text{ kWh/year per m}^2$$

$$E(50°) = (0.85 \times 0.90 \times 705) - (4.5 \times 46) = 332 \text{ kWh/year per m}^2$$

Thus heat production from a 1 m² solar collector is reduced by just over 40 kWh if the working temperature during the year is raised by 10°C on average. In the same way it can be calculated that the heat output is increased by 32 kWh/m² if the transmission factor of the glass is raised from 85% to 90%.

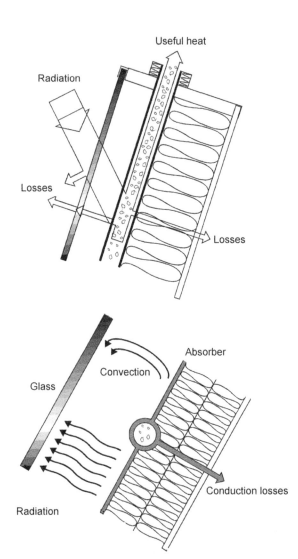

FIGURE 2.19 *Energy flow and heat losses in a solar collector. The radiation that falls on the solar collector is reflected and absorbed by the cover glass. The angle of incidence affects the optical losses. The transmittance of window glass is slightly under 90%. Many manufacturers choose tempered glass with a low iron oxide content, which increases the light transmission by about 3%. In recent years anti-reflex treated glass has often been used, which further improves the light transmission. A certain heat loss by radiation from the absorber is inevitable. With a selective surface, the emittance (re-radiation) is reduced to just a few per cent, whereas the absorption is over 90%. There are also heat losses from the solar collector as soon as the absorber temperature exceeds the ambient temperature, partly down through the insulation and partly up (in the form of both radiation and convection) through the glass*

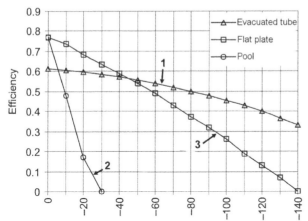

FIGURE 2.20 *The heat production from a solar collector varies with the amount of solar radiation and the size of the heat losses. For the same solar collector with the same amount of solar insolation (800 W/m² in the curves shown here), the heat output can vary with the working temperature. It is remarkable that the efficiency curve (see curve 2) for a pool solar collector is higher than for a glazed, well-insulated, flat-plate solar collector at low working temperatures. This is partly because the reflection losses are less. With a higher working temperature the efficiency is drastically reduced, which limits the usage of pool solar collectors. The efficiency of a well-insulated, flat-plate solar collector is shown in curve 3. A characteristic of a high-temperature solar collector is that efficiency is maintained at high working temperatures (curve 1)*

evacuated solar collectors have considerably better efficiency (about 640 kWh/m² per year) with greater differences in temperature, and unglazed pool solar collectors have better efficiency at lower working temperatures.

For the consumer it is the heat replacement (potential savings) that is most interesting. This is influenced by many factors: heat production from the solar collector, the system construction, and the heat that is replaced.

2.3.9 TESTING AND CERTIFYING SOLAR COLLECTORS

Since the middle of the 1990s those concerned with solar heating in Europe have worked to develop testing standards for solar collectors and solar heating systems. The work resulted in two new standards being approved

Box 2.5 Efficiency equation

SP's solar collector tests define the efficiency of the solar collector: that is, the relationship between the radiation on the collector and the net heat energy delivered by the collector. Three parameters, h_0, k_0 and k_1, are determined from measured data in a mathematical model that allows calculation of the efficiency of the solar collector at different temperatures and insolations. h_0 is given in %, k_0 in W/(m²°C) and k_1 in W/(m²°C²).

The efficiency equation is

$$h = h_0 - \frac{k_0}{E_t}(T_f - T_1) - \frac{k_1}{E_t}(T_f - T_1)^2$$

where E_t is the actual radiation in W/m²; T_f is the solar collector working temperature ($= (T_{in} + T_{out})/2$); and T_1 is the ambient air temperature. h_0 shows how much of the solar radiation reaches the absorber and is absorbed there. This should be as large as possible.

k_0 and k_1 show how large the solar collector's losses are. With the help of these the effective loss coefficient, k_e, can be calculated:

$$k_e = k_0 + k_1(T_f - T_1)$$

The lower the value of k_e, the better the solar collector is insulated.

The value of k_e for flat-plate solar collectors normally varies between 3.8 and 4.5 W/(m²°C) when $\Delta T = 50$°C. Evacuated solar collectors have a value of k_e between 0.8 and 1.5 W/(m² °C) when $\Delta T = 50$°C. (ΔT is the temperature difference control.)

The efficiency can also be shown graphically as in Figure 2.20.

Using a special version of the calculation program UMSORT, SP can calculate the solar collector's heat production per month for different temperatures. The program includes a database with hourly radiation and temperature values for Stockholm, Sweden (lat. 60° N) (1986 and 1989).

Box 2.6 Some other testing stations in Europe

ITW – Universität Stuttgart, Germany:
http://www.itw.uni-stuttgart.de
DTI – Dansk teknologiskt institut, Denmark:
http://www.solenergi.dk/inenglisch.asp
ITR Rapperswil, Switzerland: http://www.solarenergy.ch
Frauenhoferinstitutet, Germany: http://www.ise.fhg.de

by a majority of the countries participating in CEN in 2001. The new testing standards replace all national standards in this field. They are EN 12975 *Thermal solar systems and components – Solar collectors* and EN 12976 *Thermal solar systems and components – Factory made systems*. A third standard in preparation – *Thermal solar systems and components – Custom built systems* – was not considered ready for approval and has the status of a pre-standard (ENV 12977) at present. The European certification of solar heating products that is being developed at present, called the *Solar Keymark* (see below), will be based on the test results produced at the accredited testing laboratories and carried out according to EN 12975 or EN 12976. Various laboratories/stations throughout Europe already have, or are on the way to obtaining, accreditation for one or both of these standards. For more information on these testing laboratories and about the Solar Keymark see http://www.solarkeymark.org/

In Sweden, SP is responsible for testing and certifying (*P-labelling*) solar collectors. The aim of P-labelling is to provide impartial information on the properties of different makes of solar collector (see Box 2.8). For consumers, authorities and manufacturers the P-label provides assurance that a product meets specific requirements for function, durability and performance (see Box 2.9). To achieve P-labelling the manufacturer

signs an agreement to have the plant inspected annually. SP's test list is updated continually. It can be obtained from SP, but is available only in Swedish. About 15 solar collectors have been P-labelled since 1990 when the system began. The requirements for P-labelling have recently been revised, in line with EN 12975. New products and materials have also led to the revision of some requirements. However, the testing of materials, which is included in P-labelling, is outside the scope of the European standard.

SP is also working actively for a common European certification of solar heating products within the EU Altener project Solar Keymark. Keymark is a CEN/CENELEC European label, organized by the European Committee for Standardization, which works mainly for the establishment of EN standards as requested by manufacturers or the market. The Keymark label is a relatively newly established symbol, which can now be found in 19 European countries. It is a voluntary European quality label, acquired by testing according to a European standard. The aim of this common certification system is that a test carried out in one European country will be sufficient to allow the product to be sold in the whole of Europe, thus making it easier and cheaper for manufacturers to increase their markets. Increased competition leads, in turn, to better and cheaper products.

Box 2.7 Basic information on solar heating

- Solar collectors produce more heat at lower working temperatures (under the same external conditions).
- Solar collectors give varying temperatures depending on solar insolation but also depending on the temperature in the heat store.
- The heat transfer rate of the solar collectors cannot be predetermined but varies with solar insolation.
- A 10 deg C reduction in the average temperature in the solar collector generally gives 15–20% higher heat output, depending on the construction of the solar collector.

Box 2.8 SP's test of quality

Before being certified by SP (P-labelling) the solar collector must undergo a number of tests:
- examination of drawings and specification of materials
- examination of installation and maintenance instructions
- pressure tests
- stagnation and thermal shock tests
- stability for wind and snow loads
- imperviousness to rain
- definition of thermal performance
- material tests (certain components)
- preparation of report on properties.

After the tests an accelerated test is carried out: that is, an outdoor exposure for one year in a state of stagnation. If all the stipulated test requirements are fulfilled and there are procedures for future quality controls, the solar collector can be P-labelled.

Box 2.9 Quality requirements for solar collectors

- High temperatures (stagnation temperatures for a solar collector can exceed 180°C) make demands on the choice of insulation material, the construction of the box (for example the spontaneous ignition temperature of wood), the choice of absorber, and other component materials (such as rubber strips, and insulation of the box and pipes).
- The construction of the box and fixing must stand up to the wind and snow loads that can occur.
- The solar collector must stand variations in temperature. It is not unusual for the temperature in the solar collector to change from a very cold night temperature to a high working temperature in a relatively short time (a few tens of minutes).
- The solar collector must be impervious: that is, it must be able to withstand rain and damp. Too much condensation means a reduction in the solar insolation.
- The solar collector (and other material that is exposed to sunlight) must be resistant to UV radiation over a long period (30–50 years), and this applies to all the component materials.
- All the components in the system must stand up to attack by animals. Certain plastics and rubber materials, insulation etc. are attractive building materials for rodents, birds and other small animals.
- The solar collector should be easy to work with, both for installation, and for service and maintenance.

The major difference between Keymark and P-labelling is the material tests that are included in P-labelling. This means that a P-labelled product has gone through all the tests that are required for Keymark. However, there can be additional requirements and documentation, for example that the requirements for the manufacturers' own checks are a little tougher for Solar Keymark.

The annual output calculated by SP is for solar collectors facing south with a 45° tilt. The weather data used are for Stockholm in 1986 (latitude 60° N); solar insolation was then 1062 kWh/m² annually. Calculations are made using the simulation program UMSORT with SP's test results as input data. The annual output that is given should mainly be used as a means of comparison. The actual annual output varies with the design of the system (solar collector, working temperature), solar collector orientation and users' habits, as well as with

the available solar radiation. The reference area that is given (normally the solar collector's transparent aperture area) is the basis for calculating the collector's thermal performance and annual output. Larger continuous units achieve better values, mainly because the edge losses (of heat) are less per unit area. SP uses operating temperatures of 25°C, 50°C or 75°C, which refer to the average values of the inlet and outlet temperatures of the heat transfer fluid on its way through the collector. This temperature is compared with the temperature of the collector's surroundings.

2.4 The solar circuit

2.4.1 COMPONENT PARTS

This section describes the various components in the solar loop. The descriptions are based on Swedish conditions: for example, the severity of the climate necessitates the use of a freeze-resistant heat transfer fluid, which in turn makes special demands on the design of the system and component parts. Solar heating has a simple system technology with few components. This means that costs can be kept down, and that operation and maintenance work is made easier. The components described here refer mainly to systems for single-family houses.

2.4.2 OPERATING PACKAGE

In recent years an important technology development, known as the operating package, has been developed for the single-family house market. It contains all the essential components for the solar heating system, including the pump, control unit, expansion vessel, filling and drainage valves, plus safety valve, filters, control valve, non-return valve, pressure gauge, temperature sensors and, in some cases, a flowmeter (Figure 2.21).

A short description of the most common and frequently used components in a conventional pumped, pressurized solar heating system is given below.

2.4.3 PUMP

The pump is one of the most important parts of a solar heating system; it is the real heart of the system (Figure 2.22). It transfers the heat content of the solar radiation from the collector to the heat store. A control unit starts the circulation in the solar loop as soon as the temperature in the collector exceeds the temperature at a reference point in the storage tank (heat store). The pumps are relatively small compared with the amount of heat that is transported. In a normal system for a single-family house in Sweden the power varies from 30 W to 80 W (approximately 90 kWh of running energy is used to obtain up to 4000 kWh of solar heat). The annual running costs in this example are around 60 SKr (€6.67).

When sizing the pump, it is important to determine the flow rate in a solar heating system first, as the volume flow rate (litres per minute) affects the pressure drop (see Figure 2.23). The estimated pressure drop gives the capacity of the pump. It is complicated (expensive) to increase the flow in an existing plant because the pressure drop increases more quickly than the flow rate. The supplier of the solar heating system should recommend both pipe dimensions (see Box 2.10) and the capacity of the pump. When calculating the pressure drop, it is the heat transfer fluid that is the determining factor, not the temperature variation in the circuit. For example, the drop for a glycol–water mixture will be far greater than for water alone. Sizing is for a given temperature level and a given flow rate.

The pump chosen must be suitable for the heat transfer fluid used. There are various different makes of solar heating system that use glycol mixed with water. If solar brine or another heat transfer fluid is used this makes demands on sealants and other materials, both in the pump and for the solar loop. Always check with the supplier and manufacturer and/or the rules and regulations in force in the relevant country.

It is also important that the construction of the pump can stand frequent starts and stops. One interesting type of hot water circulation pump is constructed without an axle, which means that there is no connection between the winding and the turbine. The turbine is made of nylon and the pump housing of

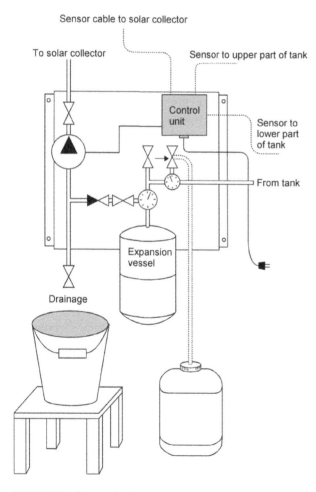

Sensor cable to solar collector

To solar collector

Sensor to upper part of tank

Control unit

Sensor to lower part of tank

From tank

Expansion vessel

Drainage

FIGURE 2.21 *Some solar heating suppliers have developed complete system packages. This includes most, or all, of the important parts of the system. Note the container to collect the heat transfer fluid from the safety valve if there is excess pressure*

Box 2.10 Calculation of volumes in pipe system

Cu pipe:
10 × 0.8 mm 0.055 l/m (litres per metre)
12 × 1.0 mm 0.078 l/m
15 × 1.0 mm 0.133 l/m
18 × 1.0 mm 0.201 l/m
22 × 1.0 mm 0.314 l/m
28 × 1.2 mm 0.515 l/m

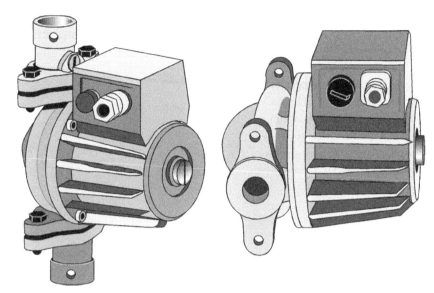

FIGURE 2.22 *Pumps in the solar loop are relatively small. The pump in a single-family house system has the same power (30–100 W) as a normal light bulb. The height rise (pressure drop) normally determines the size. Most makes of pump have three or four different choices of speed*

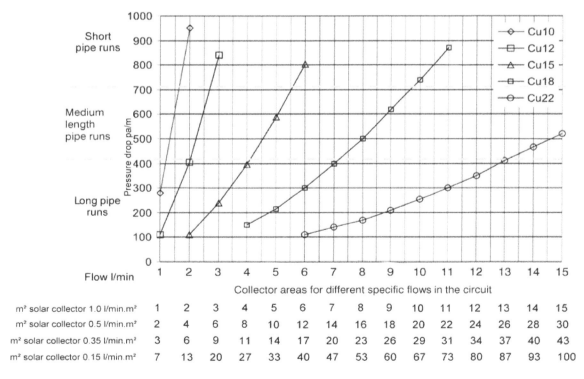

m² solar collector 1.0 l/min.m²	1	2	3	4	5	6	7	8	9	10	11	12	13	14	15
m² solar collector 0.5 l/min.m²	2	4	6	8	10	12	14	16	18	20	22	24	26	28	30
m² solar collector 0.35 l/min.m²	3	6	9	11	14	17	20	23	26	29	31	34	37	40	43
m² solar collector 0.15 l/min.m²	7	13	20	27	33	40	47	53	60	67	73	80	87	93	100

FIGURE 2.23 *Pressure drop for copper pipes, dependent on flow and dimensions of pipes for 35% propylene glycol at 40°C. The pressure drop and the flow for different conditions can be read from the diagram. The pressure drop is given for the pipes alone, not for the collector as a whole*

bronze. There are a number of standard pumps that are suitable for the heat transfer fluid used.

There can be vibration sounds from the pump, and these are transmitted and 'follow' the pipe run. The sound can be amplified by vibration or by resonance in the spaces along the pipe run. This sound can be avoided if the suspension mounting of the pump is 'devibrated' (for example with the help of a flexible metal tube on each side), and if the pipes are drawn through spaces where pump noise does not disturb and is not amplified. Pump mountings with built-in stop valves are recommended to facilitate servicing and maintenance.

2.4.4 CONTROL VALVE AND FLOWMETER
Control valves can be used to ensure a uniform flow through all the solar collectors. In larger systems a control valve is fitted for each group of solar collectors. A simple flowmeter can be useful to check the rate of flow. Adjustment of the flow in smaller systems is by choice of pump speed unless otherwise specified.

2.4.5 NON-RETURN VALVE
To prevent unplanned natural convection (which would result in a temperature drop) a non-return valve should always be fitted to a solar heating system. (See Box 2.11 for a brief explanation of natural convection.)

In some cases the natural circulation is so strong that a spring-loaded non-return valve cannot withstand the forces. In these cases a magnetic valve governed by the control unit is recommended. Fitting and control can vary in different makes.

When choosing non-return valves it is important that the materials in them will not be corroded by the fluid used in the system, or damaged by the high running temperatures that can occur. It is advisable to install shut-off valves on either side of the non-return valve or magnetic valve.

2.4.6 FILTER
All sorts of dirt and particles can accumulate in the solar circuit's heat transfer fluid and cause problems, partly because they can damage the component parts (pump, flat-plate heat exchanger, non-return valves) and partly because they can affect the circulation (flow rate).

> ## Box 2.11 Natural convection
> Natural convection starts with differences in density in the heat transfer fluid. Higher temperatures give lower densities: the fluid becomes lighter and rises. The surest way of ascertaining whether there is natural convection is to measure the differences in temperature on the pipe runs to and from the solar collectors. The measurement should preferably take place under extreme conditions – for example, when the temperature is high in the storage tank and the solar collectors are cold (at night). A sure way of finding out whether there is natural convection is to check whether the temperature reading corresponds to the ambient temperature at night.

A filter prevents this type of disruption of operation. It must be possible to clean the filter regularly, and shut-off valves should therefore be placed on either side of the filter so that it is not necessary to drain the solar loop for cleaning. When the filter has been cleaned it is important that the pressure is restored in the system and that air is bled off the circuit. The frequency and method of cleaning the filter should be included in the running and maintenance manual.

2.4.7 FILLING AND DRAINAGE
The solar loop is filled and drained by the same valve, which should be placed at the lowest point of the loop. In larger systems it is normal for a filler vessel (which is also used to ensure the stipulated pressure in the system) and a fixed filler pump to be placed next to the lowest point on the loop. The vessel is used on filling (for example as a mixing vessel for the glycol and water of the heat transfer fluid), but can also be used as a collecting vessel in the case of excess pressure. In this case an overflow pipe leads from the safety valve to the filler vessel. To make it easier to fill the solar loop a manual air bleeder can be fitted, either at the highest point of the circuit or as an automatic air bleeder on the main pipe.

2.4.8 EXPANSION VESSEL/DRAINAGE VESSEL
The increase in volume of the solar circuit's heat transfer fluid is taken up by an expansion or pressure vessel as temperatures rise (see Figure 2.24). Just as in a

conventional heating system the expansion vessel should be sized according to the volume in the solar heating system's closed circuit. The size of the expansion volume is determined by the total volume in the solar loop and the temperatures that occur (which can vary considerably in a solar heating system, from –30 to +180°C). It is important that the quality of the vessel is suitable for the type of heat transfer fluid used and the high working temperatures that may occur. Normally the manufacturers provide a pressure vessel that is standard for the respective make. The rules and regulations for the relevant country must also be followed.

There is normally a manometer, a safety valve (which is always required in a closed circuit), and an overflow pipe to a separate collecting vessel next to the expansion vessel. It is important that there is some means of expansion in the solar collectors: in other words there must not be any means of shutting off the connection between the solar collectors and the safety valve (expansion vessel). Always follow the supplier's instructions and/or the regulations in force in the relevant country.

The boiling point of the solar circuit depends on the pressure in the system and the properties of the heat transfer fluid. Normally the system has overheating protection in the control equipment, which stops circulation in the solar collectors at a preset temperature. The risk of steam forming in the solar collector can be avoided by raising the pressure in the

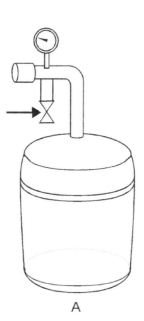

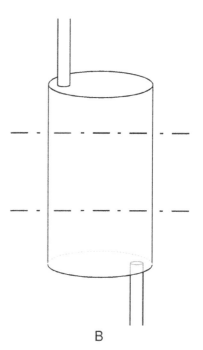

A

B

FIGURE 2.24 *Example of systems for single-family houses. The temperature of the heat transfer fluid in a solar heating system varies greatly. At some times of year the difference in temperature from night to day can be very large. The temperature in the solar collector can rise by 50–60°C in just a few minutes. Two main principles are used to accommodate the increases in volume. (A) Pressure vessel used in the system using glycol and water, which is by far the most common type. A rubber bladder in the vessel takes up the increase in volume. A safety valve is normally placed adjacent to the manometer. It is important to provide a separate overflow pipe from the safety valve to the collection vessel, partly because the glycol mixture is toxic, and partly because it is environmentally damaging. (B) In a drainback system a drainage vessel is used to collect the heat transfer fluid from the solar collectors when they are not in operation. The drainage vessel also takes up the increase in volume when the working temperature increases in the solar loop. Even a drainback system must have a safety valve to remove possible excess pressure in the loop. If the heat transfer fluid is not freeze-resistant the drainage vessel must be in a place free from frost*

system or by draining the solar collectors. Recurrent problems of overheating indicate some sort of fault in the system.

If glycol mixed with water is used as the heat transfer fluid it is advisable to have a pressure vessel. By increasing the working pressure in the solar circuit to 4–9 bar (particularly in smaller systems, such as for the single-family house market) the boiling point is raised to 140–180°C, which gives greater scope for eliminating problems of overheating. The stagnation temperature (maximum temperature) of the solar collectors affects the choice of working pressure and the pressure at which the safety valve opens. A well-insulated flat-plate collector has a stagnation temperature that can reach over 180°C at full solar insolation without cooling (circulation) of the solar collector. The stagnation temperature for evacuated solar collectors is even higher. Note that pressurized systems make special demands on component parts and installation: follow instructions and safety regulations. In larger solar heating systems, such as district heating plants, a pressure vessel with constant pressure can be used.

In drainback systems a drainage vessel is placed below the lowest point of the solar collector and kept free from frost. The drainage vessel is used to collect the volume of fluid from the solar loop, which can be exposed to temperatures below 0°C or excess temperatures. Although a drainback solar heating system is pressureless, a safety valve is normally placed on the loop. One of the advantages of drainback systems is that ordinary water can be used as the heat transfer fluid.

2.4.9 CONTROL AND SUPERVISION

The control unit of a solar heating system starts the circulation in the solar loop when the temperature is higher in the solar collectors than at the reference point in the heat store. Normally some sort of ΔT control (temperature difference control) is used with two temperature sensors, one placed in the solar collectors and the other at the inlet to the heat store. The difference in temperature is normally preset to 5°C to start and 2°C to stop circulation. Most automatic solar

heating control units for the single-family house market have a temperature display where the temperature of the sensors can be read. A further temperature can be read using a third temperature sensor, for example in the upper part of the storage tank. The ΔT control often has built-in overheating protection, which stops the circulation in the solar loop when the heat store has reached a preset temperature. The overheating protection device can, for example, be set at 28°C for an outdoor pool or 95°C for a storage tank. Figure 2.25

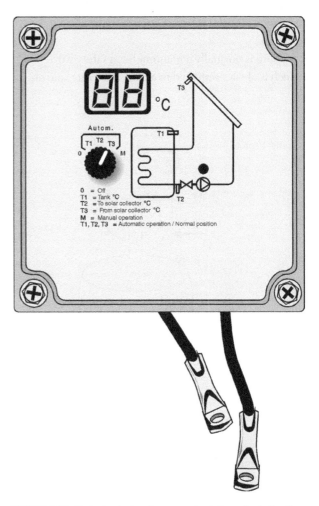

FIGURE 2.25 *Typical solar heating control unit for a single-family house system. The unit has a display to show the temperatures in the solar collector and the tank. Overheating protection is normally included*

shows a typical heating control unit for a single-family house system.

In larger solar heating plants the control equipment starts circulation at a preset temperature, given by a separate reference solar collector. This sort of control is appropriate when there is a large volume in the solar loop. In a solar heating plant for district heating, with long culverts between the solar collector and the storage tank, initial internal circulation of the heat transfer fluid through the solar collector loop can be justified. In this case heat transfer into the storage tank is not started until the rise in temperature is sufficient in the whole loop (this also means that there is no need for a long cable between the solar collector field and the sensor in the control room). This is called a *two-stage start-up*: the solar loop starts in the first stage, with a signal from a reference solar collector or at a preset start temperature in the control unit, and the loading loop in the second stage. An alternative to a secondary loop is a three-way valve that opens the inlet to the heat store at the required temperature difference.

In the same way a start at a preset temperature can be a simpler and more reliable control technique than an ordinary ΔT control for pool plants.

The position of the sensor is extremely important for the control equipment to work correctly. The sensor on the solar collectors is subject to great stress. The equipment works best if the sensor can be mounted directly in the solar collector, for example in a ready-installed immersion tube. It is important to place the sensor as near the solar collector as possible so that the heat can be led directly to the temperature sensor. It is also important for the sensor to be well insulated from the ambient temperature and the wind cooling effect. Use insulating material that will withstand the high temperatures that can occur near the solar collector. Rodents and birds can do damage, and this affects the choice of material and the installation.

The sensor in the heat store must register the 'right' temperature: that is, the temperature at the level where the solar heat will be delivered. In a stratified storage tank the temperature sensor should be placed at the level of the heat exchanger. With an alternative position outside the heat store, attention must be paid to possible losses from the pipes and influence from the ambient temperature in the room. Place the sensor as near the heat store as possible and insulate it well.

Disruption in the operation of the control unit is normally caused by a fault in the sensor, either because the sensor unit has not stood the temperature load or because the sensor cable has been subject to outer damage. With an ohm-meter (in the case of a resistance sensor) it is easy to check the function of the sensor. On delivery it is important to enclose a technical specification of the control unit with a description of operation.

Some suppliers can offer more advanced control equipment with a variable-speed pump and accurate documentation of running times and, in some cases, a recording of heat production from the solar collectors.

Larger solar heating plants are often equipped with a central alarm system, which alerts emergency personnel if there is a breakdown. Alarm systems of this type can be set off when the pumps in the plant break down or if the system is affected by problems of excessive temperatures.

2.4.10 HEAT TRANSFER FLUID

Glycol (normally propylene glycol with an inhibitor) mixed with water is the heat transfer fluid that is normally used in solar heating plants, at least in the northern parts of Europe. However, water on its own may be common as a heat transfer fluid in the future, with the development of drainback systems. In direct solar heating systems for outdoor pools chlorinated water is used as the heat transfer fluid (see Chapter 6).

In recent years a special brine solution has been tested as the heat transfer fluid in small-scale solar heating systems. This solution, like glycol, consists of organic carbon–hydrogen chains. It has good heat transfer properties and is biodegradable. One of its disadvantages is that it tends to leak (particularly in combination with tow sealing); it is too soon to say anything about brine's ageing, performance and long-term properties in a solar heating system.

There are some problems and risks with glycol in solar heating systems:

■ It has been shown that the freezing point can change with time or if the glycol mixture is subject to temperature shocks. For this reason it is important to check the freezing point of the glycol mixture regularly, for example before each winter season.

■ Propylene glycol mixed with corrosive water (low pH) can damage the main pipes in the solar heating loop. Take samples at regular intervals (every third to fifth year) for chemical analysis to establish the corrosiveness of the glycol mixture.

■ The boiling point is another problem. To raise the boiling point the solar heating system can be pressurized. Check that all the components will stand this pressure, and that it is permitted. An overflow pipe from the safety valve on the solar loop to a collecting vessel is recommended. This is so that the glycol mixture does not run out in the sewer or damage anything in the control room. This should be taken into consideration when positioning the pressure vessel, safety valve and collecting vessel.

2.5 Heat exchangers

A heat exchanger is used in solar heating systems in which the heat transfer fluid circulates in a closed loop. The heat exchanger transfers the heat from the fluid in the solar loop to the heat store without mixing the fluids. In direct solar heating systems (such as thermosiphon systems or pool plants) the same fluid is used for heat storage and heat transfer: thus no heat exchanger is needed.

A suitable type of heat exchanger is chosen based on the properties of the solar heating system. The choice of heat exchanger and its size are determined by the size of the solar heating system, the volume of the storage tank, the flow in the solar loop, the temperature intervals and the type of heat transfer fluid. The most normal types are either copper helical finned-coil heat exchangers or stainless steel flat-plate heat exchangers. Small spiral tube heat exchangers are being developed and are likely to be introduced on the market shortly.

A special tube-in-tube heat exchanger has been developed for existing hot water tanks. It consists of double pipes in which the inner pipe contains the solar loop heat transfer fluid, and fresh water circulates in the outer pipe with the help of natural convection (see Figure 2.26).

2.5.1 FINNED-COIL HEAT EXCHANGER

For small and medium-sized solar heating systems the normal heat exchangers used in Sweden are finned coils, which are installed directly in the storage tank. This type of heat exchanger is frequently used in single-family house systems. There is a wide choice of standard tanks with factory-installed finned-coil heat exchangers. A rule of thumb is that the surface area of the heat exchanger should correspond to at least 20–25% of the solar collector area. This means that a heat exchanger of about 2.5 m^2 is needed for a solar collector area of 10 m^2. It is important that the finned-coil heat exchanger is large enough and is capable of transferring the solar heat from the solar loop to the heat store. It is also important that the finned-coil heat exchanger is placed correctly in the storage tank so that temperature stratification in the tank is promoted and solar heating can work at as low a temperature as possible. Always try to place the heat exchanger low down. Temperature stratification can be improved by installing double heat exchangers for DHW.

2.5.2 FLAT-PLATE HEAT EXCHANGER

In larger solar heating systems (for example for district heating plants), and for storage tanks without finned-coil heat exchangers, external flat-plate heat exchangers are used. They consist of two closed loops that prevent contact between the different heat transfer fluids, one for solar heat and one for the heat store fluids. Therefore two pumps are needed to transfer the heat content from one loop to the other. Both pumps are normally controlled by the solar heating system's control unit.

Heat transfer in a flat-plate heat exchanger is normally more efficient than in a finned-coil heat exchanger. Therefore 0.02–0.05 m^2 heat transfer area per m^2 of solar collector is normally recommended.

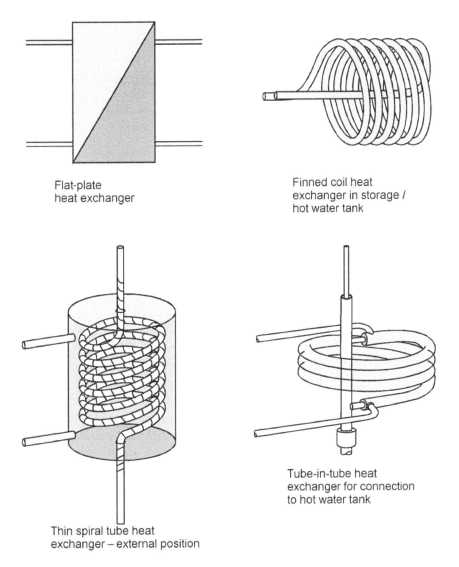

Flat-plate
heat exchanger

Finned coil heat
exchanger in storage /
hot water tank

Thin spiral tube heat
exchanger – external position

Tube-in-tube heat
exchanger for connection
to hot water tank

FIGURE 2.26 *Different types of heat exchanger*

Calculation programs for sizing flat-plate heat exchangers are available from the suppliers, for example to vary temperature demands. Flat-plate heat exchangers are normally manufactured from stainless steel (e.g. SIS 2347); the connections are of carbon steel, which requires care with oxygen-rich water. The soldering material is normally copper (99.9%).

It is important for the performance of the flat-plate heat exchanger to be considered in relation to the working conditions in the solar loop. Observe working temperatures, flow and pressure drops in both the solar loop and the load circuit. In fact it is the temperature demand and the flow in the loops that determine the area of the heat exchanger. It is also important for the heat exchanger to have a long thermal length (the thermal length of the heat exchanger describes the characteristics of the heat exchanger with regard to size and temperature differences on both sides of the exchanger). Always follow the instructions from the supplier for sizing and combinations of materials, and the relevant rules and regulations of the country.

2.6 Heat stores

The heat store is a core component of a solar heating system. Solar heat must be stored from the time of production until it is to be used. The production from the solar collector takes place when solar insolation is sufficient, and not when required by the consumer. The heat store is also essential to allow solar heat to be combined with other types of energy and, by its temperature stratification, it determines, to a large extent, the output (heat production) from the solar collectors.

Seasonal storage means that solar heat is stored from the summer season for winter consumption. With this type of heat storage it is completely realistic for the solar fraction to reach 60–75% of the total heat load, even in northern Europe.

A short-term store is used as a heat store from the time of production to the time of consumption. Thermal short-term stores are used mainly to smooth the load in district and local heating networks. A short-term store

---

also allows the use of differentiated tariffs (day, night and weekend tariffs). At present researchers and manufacturers agree that the solar heating plants that have the best competitive basis are those with short-term storage. The traditional storage technique uses steel tanks that are connected to the heat production system.

Figure 2.27 shows three different types of heat store.

2.6.1 STORAGE SYSTEM

With the present uncertainty over future energy prices it is not difficult to make a case for flexible heating systems. A central component in this context is the storage tank, which allows the optimal production and distribution of heat in free combination with other sources.

A storage tank system is dimensioned with regard to the heat sources included in the system and the extent to which they need a heat store to safeguard the heat load and work well. There are in fact few heat producers that need a heat store. An important exception, however, is wood-fired boilers for single-family houses. These normally have a much larger heat output (combustion chamber) than the average heat requirement of the house. For safe and easy combustion a heat storage tank is used to store the heat from the time of combustion until the system requires heat. In this type of system there is a volume conflict between the wood-burning equipment and solar heating. This can result in a storage volume divided into several units: a master tank (with solar coil, used for domestic hot water production) and a slave tank that is used only to increase the volume when the wood-fired boiler is in use. The volume of the master tank should be 50–100 litres per m² of solar collector area (75 litres per m² of solar collector area is a good guideline) depending on the design of the system. Too small a tank volume in relation to the solar collector reduces the solar collector's efficiency, as its working temperature rises quickly, which leads to increased heat losses. Small storage volumes also have less heat storage capacity. Large storage volumes result in a slow temperature rise. The solar collector's efficiency is not negatively affected (indeed, rather the opposite), but it takes a long time for the tank temperature to reach consumption

---

> ## Box 2.12 Storage of solar heat
>
> Storage of solar heat is either in a seasonal store, in which the heat produced in the summer is stored for winter consumption, or in a short-term store, which keeps the solar heat from when it is produced until it is to be consumed, or forms a buffer between sunny and less sunny days.
>
> Seasonal stores for larger solar heating projects should be able to store about half of the annual heat demand, which can mean an annual coverage of between 60% and 75% of the total heat load. Variation in the annual coverage depends on the size of the project and the form of storage, and also on the willingness to invest.
>
> Short-term stores for the single-family house market (ordinary storage tanks) are sized to cover 2–3 days' heat demand during the summer period. Solar heating can manage to store heat from sunny to cloudy days, when consideration is given to the summer load, size of solar collectors and storage tank.

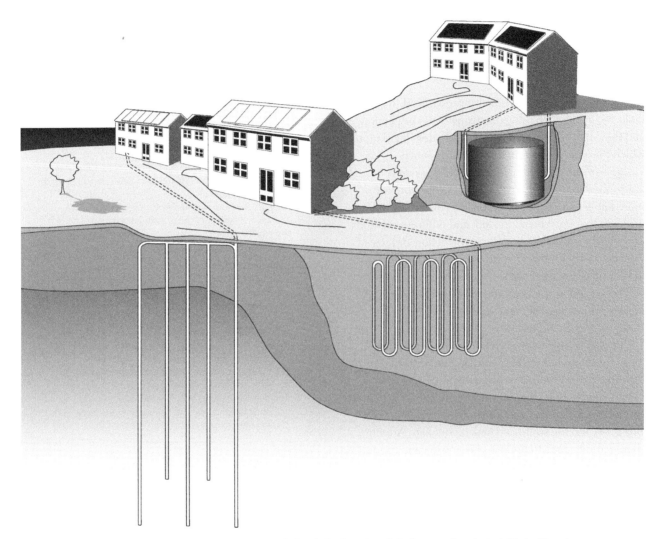

FIGURE 2.27 *Different examples of heat stores. From the left: boreholes in rock, coils in the ground, and a tank filled with water*

temperature (50–70 °C), with the risk that the share of auxiliary heat may increase.

Box 2.13 lists the selection criteria for storage tanks.

2.6.2 STEEL TANKS

Steel tanks are the most common heat storage method for solar heating systems, particularly as short-term storage for relatively high temperature ranges (Figure 2.28). The short-term store is also the exchange unit, and smooths the load between the solar loop and the distribution network.

There are three main types of steel tank:

- simple steel tanks for heat storage at atmospheric pressure, most common for the single-family house market and in multi-family dwelling projects
- pressurized steel tanks for heat storage with a pressurized system, intended for smaller district heating networks and local heating plants; there are also small pressurized storage tanks for single-family houses

■ larger steel tanks for heat storage at atmospheric pressure with a water vapour barrier as the expansion volume, a technique that is used only in larger district heating networks.

There are safety requirements for pressurized steel tanks regarding such things as strength and inspection, and they may be the subject of special safety legislation, which must be checked in the relevant country.

Non-pressurized steel tanks, which operate at atmospheric pressure, are normally cheaper but are more liable to corrosion. Oxygenation of the water is a problem that must be addressed, and different methods are available depending on the construction and size of the heat store. It is also important to allow for changes in volume of the fluid in the heat store.

Always follow the recommendations from the designer, manufacturer and supplier, and observe statutory requirements and regulations.

Steel tank technology is relatively well tried. Steel tanks for single-family houses are a standardized technology today and have achieved a commercial breakthrough. Steel tanks for district heating systems (projects in Falkenberg and Nykvarn in Sweden – see Box 2.14) also use well-tried tank technology. Attention should also be paid to the risk of corrosion and to

---

Box 2.13 Selection criteria for storage tanks

- What is the relevant main heat source (oil, gas, electricity, wood or other)?
- What is the heat load, and which type of energy is to cover the peak demands?
- How will the DHW be heated (for example in the storage tank or in an external hot water tank)?
- How will the heat be distributed (radiators, floor heating, airborne heating, etc.) and what is the required supply temperature?
- How much space is available for the storage tank?
- Design and construct the system (storage tank) so that the auxiliary heat can be turned off when the solar heating system starts.

Also remember:
- to design the tank so that the efficiency of the solar collectors is optimized through the correct size and maximum temperature stratification
- to design for maximum solar fraction so that the auxiliary heat can be turned off as soon as possible
- to utilize the rise in temperature from the solar collectors in the correct way
- to simplify the system design, which leads to safer operation and lower running and service costs
- to provide conditions for a good installation with minimum heat losses from pipes, heat exchanger and storage tank.

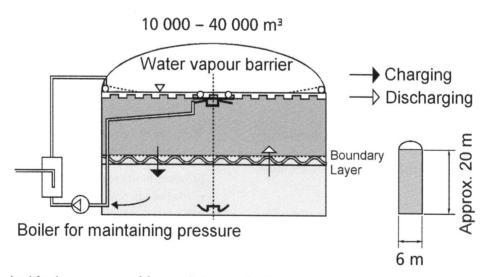

FIGURE 2.28 *Steel tank for short-term storage of, for example, hot water for district heating or local heating systems*
Source: Swedish Council for Building Research (BFR) (1994)

guaranteeing the pressure and oxygen content in these large storage tanks.

2.6.3 PIT STORES

The concept of the pit store (Figure 2.29) is that the construction should reduce heat losses and cut investment costs. The pit store is often at ground level and consists of a sealing layer with the addition of insulation. The insulated cover is normally loadbearing. The difference in density of the water at different temperatures contributes to temperature stratification of the volume of water, where the hotter (lighter) water stays in the upper part of the store. New material technology (such as butyl rubber and metal membranes) allows storage temperatures up to 95°C. Pit stores of sizes between 100 and 30,000 m³ should be common soon.

Since the beginning of the 1980s five pit stores have been built in Sweden for the seasonal storage of solar heat. Box 2.15 and Figure 2.30 describe the pit store at Åmmerberg, and Box 2.16 and Figure 2.31 describe the pit store in Malung.

2.6.4 ROCK CAVERN STORES

At present the best technique for the seasonal storage of solar heat is the use of heat stores in rock. Underground heat stores in rock, filled with water and completely uninsulated, have been tested in several projects. The technology is well known from several different areas of

---

Box 2.14 Short-term storage at Nykvarn

Telge Energi's solar collector field at Nykvarn, near Stockholm, Sweden, has a total area of 7500 m². The solar collectors are connected to a water store of 1100 m³. The heat store is a welded steel tank construction, 30 m high, insulated with 60 cm mineral wool and clad externally with aluminium. The tank can be heated to 95°C. To improve temperature stratification the heat input can be at two levels. The outgoing water to the district heating network is mixed from two different levels, depending on the temperature required. The heat production from the solar collectors on a sunny summer's day corresponds to half a week's consumption and can be stored in the steel tank. The heat store can be used during the winter – for example when the boilers are being serviced.

---

use, including oil storage in rock caverns, shelters and tunnels. In the solar heating context, competent evaluation of material and heat technology has been made of the Lyckebo project in Uppsala, Sweden (see Figure 2.32). The Lyckebo project consists of a water-filled rock cavern store with a volume of 100,000 m³ built in 1984. The results here are very good, except for deposits in the heat exchanger and the excessive heat losses (normal heat loss in a rock cavern store should be less than 10%). High construction costs have resulted in a new technology, and it is possible that existing rock caverns – disused oil storage, for example – may be more widely used in the future.

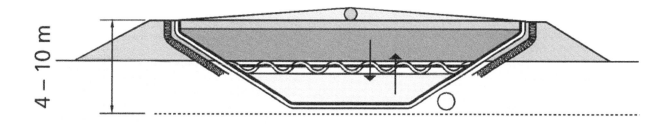

FIGURE 2.29 *Pit store in the ground, suitable volume 1000–50,000 m³*
Source: Swedish Council for Building Research (BFR) (1994)

Box 2.15 The Åmmeberg project

The Åmmeberg project is a water store based on a 'floating wall' principle (patent applied for). The technique is used to create an energy store in a cost-efficient way. The store is like an upside-down cup. The cup is constructed to the required size and is lowered into a lake, rock cavern or mine which already contains water. The opening is at the bottom, the water into which the cup is lowered can move freely, and the pressure difference can be evened out through the opening. When the water in the cup is heated it is effectively captured (in the same way as the hot air is captured in a hot air balloon). A store within a store has been created.

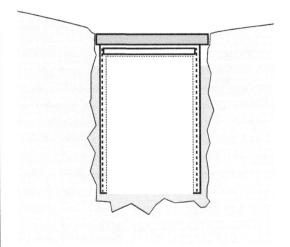

FIGURE 2.30 *The Åmmeberg project uses a pit store technique for reasonably watertight cavities based on a 'floating wall principle'.*
Source: Solsam Sunergy AB, Sweden

The cup is constructed of prefabricated polyurethane foam slabs stabilised by reinforcing iron. To stop the foam slabs from becoming wet and thereby losing a great deal of their insulating properties, they are protected on both sides by impervious layers, which in the case of the Åmmeberg project consist of thin copper sheets with soldered joints. The 'cup' is kept in position – that is, the lifting forces are restrained – by a stop being placed on the top of the cup (in this case a concrete cover). The Åmmeberg project was financed with experimental funds, and the final result shows that this pit store technique still needs further development.
Source: Solsam Sunergy AB.

Box 2.16 Pit store with thin stainless steel sheet liner in Malung

A pit heat store has been in operation in Malung since 1991. It is used as a diurnal store for storing heat from solar collectors and an electric boiler. The plant is connected to the central heating plant for space heating and DHW for about 130 dwellings. Solar collectors totalling 600m² have been designed to cover the summer load, and the main heat source is a 1 MW electric boiler.

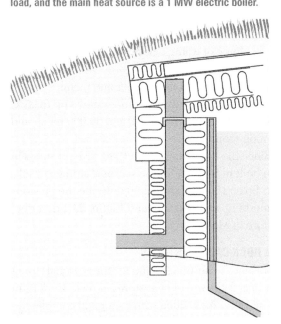

FIGURE 2.31 *The Malung project is a pit store that can be likened to an upside-down pyramid*
Source Studsvik AB (1991)

The electric boiler uses night-time electricity, as the pit store stores heat for daytime consumption.

The heat store is an insulated pit, where the ground around it consists of approximately 4 m of silt and sand. The store can be compared to an upside-down pyramid with a water volume of 130 m³. The sloping sides of the store have a layer of lightly reinforced concrete laid over a layer of sprayed polyurethane. The vertical sides of the pit are insulated with 200 mm of sprayed polyurethane insulation. The store is also insulated with Rockwool, polystyrene slabs and polyurethane slabs in the construction of the cover, which is loadbearing and designed for 400 mm insulation, 200 mm soil load and also snow load. A sealing layer of 0.5 mm stainless steel thin sheet (AINSI 316) is used in combination with 1.2 mm steel corners, which are seamed and welded on site.
Source: Studsvik AB.

2.6.5 GROUND STORAGE

Soil or clay stores are sometimes used as ground heat exchangers. The principle of the system is of interest if the working temperature of the solar collector can be kept low. This allows simpler and cheaper solar collectors to be used, and increases both the efficiency and the number of running hours of the solar heating system (during the year).

Clay storage techniques are based on the use of a bed of clay or soil as a ground heat exchanger, often in combination with a heat pump. Loosely packed soil types such as sedimentary clays or peaty soils are suitable. The solar heat is transferred to the ground store via tubes forced down vertically into the soil material (see Figure 2.33). The technology is intended for low working temperatures, normally 25–30°C. The ground store is insulated on top to reduce heat losses. The heat in the clay or soil bed can be utilized and made more efficient with the help of a heat pump that can upgrade the heat content to usable temperatures. Heat storage at temperatures higher than approximately 50°C allows a connection from the heat store direct to the heating system, without a heat pump.

With heat storage in rock using a corresponding technique, the heat is transferred to the bedrock via a large number of boreholes. This type of rock heat store is normally uninsulated and larger than corresponding storage in clay. The reason for this is that rock has a lower heat capacity and increased heat losses as its thermal conductivity is higher than that of clay. The

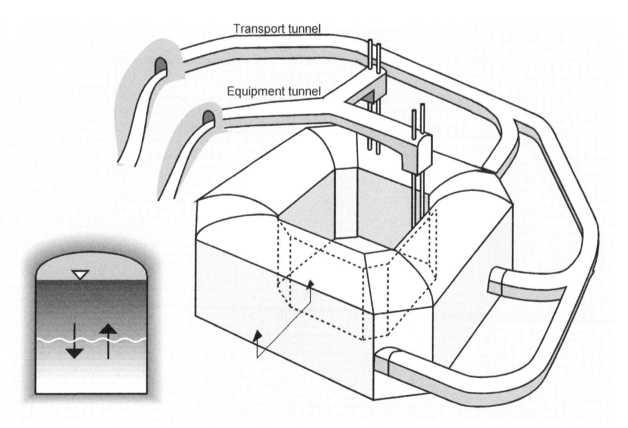

FIGURE 2.32 *Rock cavern for seasonal storage of solar heat (sketch of the Lyckebo project near Uppsala, storage volume 100,000 m³)*
Source: Swedish Council for Building Research (BFR) (1994)

higher thermal conductivity is favourable from the point of view of heat transfer, which is advantageous when positioning ground heat exchangers.

Research on this type of storage technology is being continued, for example at the Swedish Geotechnical Institute (SGI) in Linköping. A conclusion from current plants is that they work well, and that research should be concentrated on reducing the storage costs and at the same time improving the efficiency of the heat transfer between the earth or rock and the pipe system.

2.6.6 SALT STORES/CHEMICAL HEAT PUMPS

As early as 1977 a research team at KTH (the Royal Institute of Technology, Stockholm) realized that it was possible to store heat with the help of a chemical heat pump using sodium sulphide and water as active substances.

The principle of the salt storage system is to dry salt with the help of a heat source (such as a solar collector), which means that heat is stored. When water vapour is added, the same amount of heat energy as was used to dry the salts is released. The heat content can, in principle, be stored for any length of time in the salt, and it is not temperature dependent; it is necessary only to keep the salt dry. The heat is stored in the form of chemical energy (Figure 2.34).

Researchers have solved the packaging problem and the ability to transfer the heat content from the salts quickly. Great efforts are now being made to find financial backers for the start of large-scale production. Salt storage technology presents a great opportunity to break new ground in terms of costs and capacity, and at the same time give a completely new dimension and opportunity for solar heating in both large and small systems.

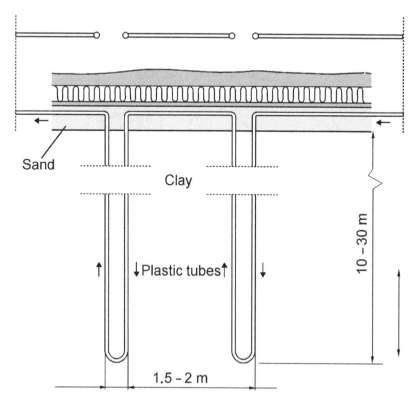

FIGURE 2.33 *Solar heating stored in the ground. Vertical U pipes in clay with extra insulation can be used together with a heat pump*
Source: Swedish Council for Building Research (BFR) (1994)

2.7 System technology

Modern solar heating technology is considered to be well developed. Many system solutions have been tried out, tested and validated by impartial institutes. Knowledge and experience of the system technology has been built up over many years. Throughout the world the aim has been to utilize simpler technology when developing solar collector construction and systems. For example, better use of materials has given both cheaper and more efficient solar collectors, and new system technology has simplified pipe installation and the component content.

Box 2.17 sets out the basic requirements to be met when choosing a heating system.

2.7.1 SYSTEM PRINCIPLES

Different system solutions are generally used for different applications; some countries also have individual solutions. In countries where the temperature falls below freezing the solar heating system has to be protected from frost, which necessitates closed solar loops with freeze-resistant heat transfer fluid circulating with the help of a pump. In more southerly countries with warmer climates, the solar heating system can be based on natural convection techniques, in which fresh water is used as the heat transfer fluid in the solar collector.

In *direct systems* the water to be heated is used as the heat transfer fluid: for example chlorinated pool water that passes directly through the solar collector. In

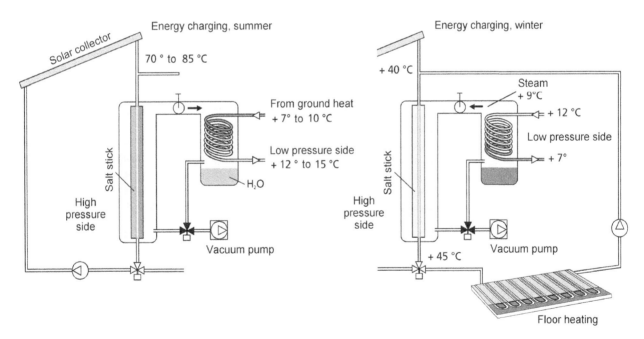

FIGURE 2.34 *Solar heat stored in salts/chemical heat pump. Heat from the solar collector heats the salts (sodium sulphide, Na_2S) in the salt store to approximately 80°C. At this temperature water vapour is released from the salts and condenses in a condensing chamber on the low-pressure side of the system. One third of the solar energy charges the salt stick, and two thirds is stored in the ground via a ground heat exchanger. When all the water vapour has been released from the salts the valve between the high-pressure and low-pressure sides is shut. The heat is now stored in the salt stick as long as it is kept in a vacuum. With discharge from the store the process is reversed. The water is evaporated and is transferred to the salt side again. The vapour is taken up in the salt store and heat is given off. One third of the energy is taken from the salt store and two thirds from the ground heat store (rock or a soil coil). The sun is the driving force. The system is being developed, and is not yet commercially available*
Source: LESAB, Fagersta, Sweden

Box 2.17 Specification of requirements when choosing a heating system

- What level of convenience is required when choosing the heating system?
- Which kinds of energy should be included?
- What is the heat load, and does it make special demands on the choice of heating system?
- If solid fuel is being considered, is there enough space (for wood-chopping, drying and storage)?
- Does the heat load of the property make special demands on the solid fuel (for storage space, for example)?
- What is the profitability requirement and the willingness to invest? Are the capital costs or the running costs decisive?
- How important is it for the heating system to be flexible?
- Are there limitations in the space available for storage tank, boilers, pipe runs, roof space for the solar collectors?
- Are there any requirements for idealistic reasons – for example that a company wishes to have a good environmental image?
- Are there any local or central government restrictions – for example on wood combustion within a built-up area, on the choice of other types of energy in a district heating area, or on aesthetical changes in the facade in connection with the installation of solar collectors?
- What are the requirements for running and maintenance? For example, should it be extremely easily maintained? Should it be inexpensive?
- Are there any requirements for simplicity and comprehensibility regarding the construction of the heating system, and what are the requirements for availability of spare parts and service personnel?
- Are there any special requirements for the choice of the type of energy – for example because of worries about expected changes in price, environmental fees, supply, or any rules or regulations that must be followed?

domestic hot water systems with natural circulation the fresh water can be used as the heat transfer fluid. In this type of system the absorber pipe must be suitable for the heat transfer fluid that is used. It is important to check the whole design of the system in direct systems so that all the component parts can stand the stress that may occur (such as quality of water, pressure and temperature).

Some direct systems are open to the surrounding air via an open expansion vessel placed at the highest point of the solar loop, or (as in a pool) direct via the pool water.

In cases where direct solar heating systems with non-freeze-resistant heat transfer fluid are used in countries with cold climates, the solar collector and pipes have to be drained in the winter to avoid frost damage. The advantages of direct systems are a relatively simple construction and a system technology that gives simpler installation work. At the same time the temperature losses at the heat exchanger are avoided. Direct systems are used mainly for pool heating. There is also an area of use in DHW systems where there is no risk of freezing: in summer cottages, for example.

More common are *indirect solar heating systems*, in which a heat exchanger is used to transfer the heat produced in the solar loop to the storage tank. These systems are normally closed and are pressurized. By increasing the working pressure, the boiling point in the heat transfer fluid is raised and, moreover, heat transfer in the system is improved. A closed solar system must be provided with an expansion vessel, a safety valve and a manometer showing the pressure. In many cases the climate demands system solutions with freeze-resistant heat transfer fluids, normally glycol mixed with water. The heat transfer fluid circulates in a closed loop with a separate expansion vessel. The heat is transferred either via a heat exchanger inside the storage tank or via an external flat-plate heat exchanger. An indirect solar heating system is always in the operational mode and can make use of all the solar insolation available during the year. An advantage of the indirect solar heating system is that the solar collector and components in the circuit are protected as they are not exposed to corrosive, oxygen-rich water.

2.7.2 NATURAL CONVECTION SYSTEMS

A natural convection system is based on the fact that hotter water is less dense, and therefore lighter, than cold water (Figure 2.35 (A)). The heat store is placed higher than the solar collectors. The cold water 'sinks' by its own weight down in the solar collectors from the heat store. When the water is heated in the solar collector it becomes lighter and pushes the heavier and colder water away. Natural convection has started and

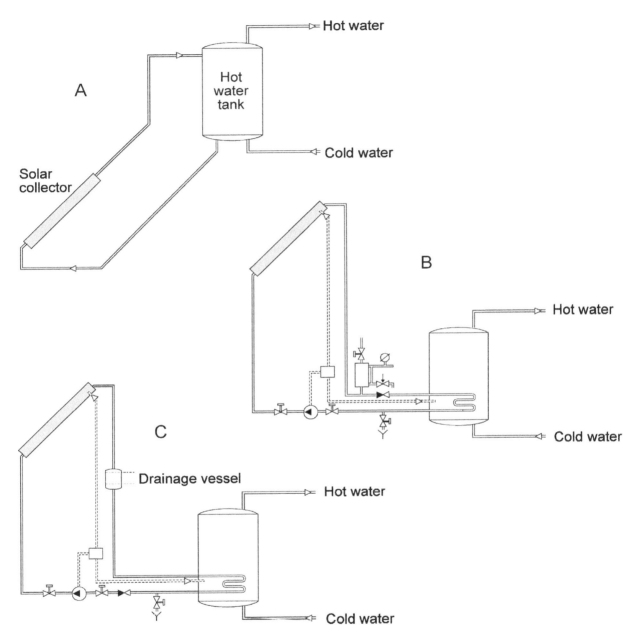

FIGURE 2.35 *(A) Natural convection system. In this direct system the same water is used in the solar collectors as in the heat store (for example a hot water tank). This type of system does not need a heat exchanger, pump or control unit. Natural convection systems are common in the Mediterranean area but can only be used during the summer in Northern Europe. It is important to ensure that there is no growth of bacteria in the hot water. (B) Pumped, pressurized system. The most common type of solar heating system in Sweden is one controlled by a separate control unit and driven by a pump. The heat is transferred via a heat exchanger in the lower part of the tank, and the increase in volume is taken up by a separate expansion vessel. (C) Drainback system. In this system the pump forces the heat transfer fluid up through the solar collector. When the control unit turns the pump off the heat transfer fluid is collected in the drainage vessel. The advantage of this type of system is that fresh water can be used as the heat transfer fluid. The drainage vessel must be placed free from frost. Carefully follow the instructions for installation from the supplier*

keeps on as long as there is a temperature difference between the solar collector and the storage tank (heat store). The system is self-regulating: the more the solar insolation heats the water in the solar collector, the greater will be the flow in the solar loop. Neither control equipment nor pump are necessary. When solar insolation is sufficient for the hot water requirements, auxiliary heat is not needed either as long as it can be guaranteed that the growth of legionella or other unwanted bacteria can be avoided. A natural convection system can work completely without electricity, and works very well in areas where there is no electricity from the grid. Moreover the system solution is reliable, simple and easy to maintain. There are complete thermosiphon system packages based on natural convection in which an insulated water container is placed higher than the solar collector (Figure 2.38).

In comparison with pumped circulation, natural convection is relatively weak. It is important that the solar collector and storage tank are near one another. The pipe run between them should be of relatively large diameter pipes and rise gently throughout its length. The solar collector absorbers must also be connected so that natural convection through the solar collector is possible.

Natural convection systems are common in Mediterranean regions. Most of the solar collectors installed annually in Greece (approximately 180,000 m^2 in 2001) are in thermosiphon systems.

2.7.3 PUMPED SYSTEMS

Solar heating systems with pumped circulation dominate the Northern European market because the climate demands freeze-resistant systems (Figure 2.35 (B)). Another important reason is that the storage tank can be freely placed in relation to the solar collector. The system technology is relatively simple. A control unit controls the pump, which starts as soon as the temperature is higher in the solar collectors than at the reference point in the heat store. Circulation continues until the temperature difference is levelled out. Pumped systems can be used in both direct and indirect system designs and in pressurized as well as open (e.g. drainback) systems.

Conventional solar heating systems for single-family houses are based on flows that vary between 0.4 and 0.6 litres/min per m^2 of solar collector area. The rise in temperature in the solar loop varies between 5°C and 20°C depending on factors such as the intensity of the solar radiation, the temperature of the storage tank, the solar collector's performance, and the flow in the solar loop. The normal flow in large systems (greater than 100 m^2) varies between 0.35 and 1.0 litres/min per m^2 of solar collector.

An increasing number of researchers and manufacturers in the solar energy industry are convinced that the future lies in *low-flow systems* (Figure 2.39). As the name suggests, a lower flow is used in the solar loop: 0.15–0.20 litres/min per m^2 of solar collector area is sufficient. Pipes with smaller diameters can be used. Conventional solar heating systems for single-family houses use pipes with a diameter of 15–22 mm; a diameter of 6–8 mm is sufficient in low-flow systems. Pipe dimensions vary with the distance between the solar collector and the storage tank, with the solar collector area, and with the manufacturer. Always follow the advice and instructions on pipe dimensions from the supplier and the rules and regulations in force in the country.

Smaller pipe diameters give lower material costs, simpler and cheaper installation and smaller heat losses. A low-flow system has a considerably higher temperature rise in the solar loop than a conventional system. It is important for good operation that the temperature stratification of the heat store be as large as possible and that the solar heat can be charged at a suitable level.

Not all solar collector models can be used in low-flow systems. One requirement is that they should have absorbers connected in series (see Figure 2.5).

One of the advantages of a low-flow system is that heat transfer can be more efficient. The solar fraction is raised, which means that the amount of auxiliary heat can be kept down. Material costs are generally lower. One disadvantage is that the control equipment is more complicated. To work well the pump should have variable speed control (even if there is constant low flow

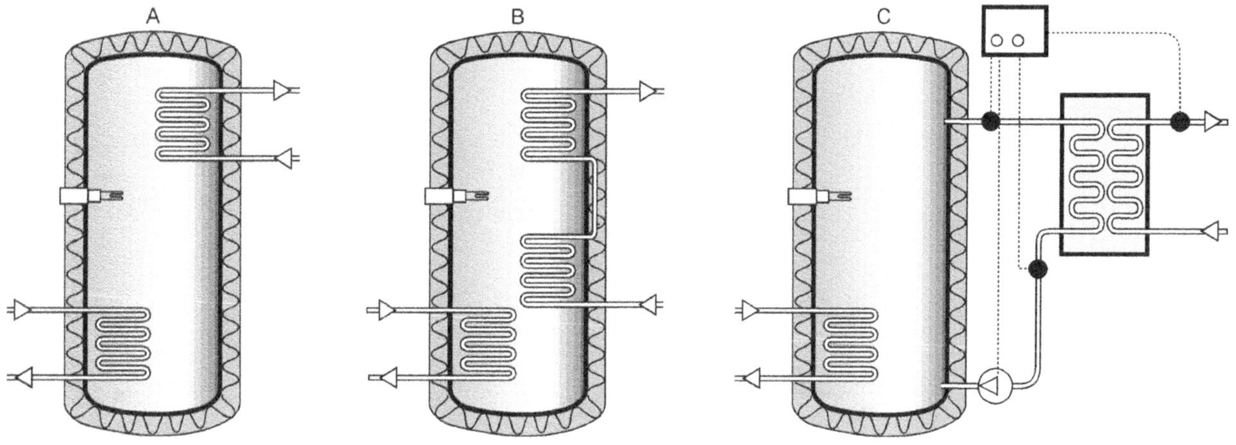

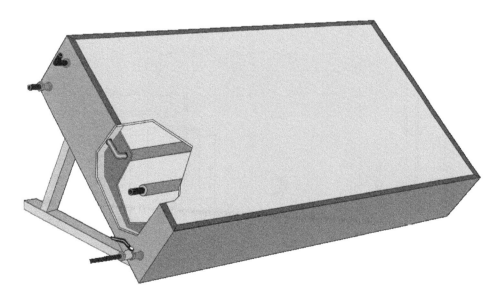

FIGURE 2.36 *(A) Internal heat exchanger for DHW. The solar fraction is low: under 40%. (B) Double heat exchangers for DHW give far higher solar fraction: almost 70%. (C) Tank in which the DHW is heated in an external flat-plate heat exchanger. This reinforces temperature stratification in the tank, giving a solar fraction of approximately 80%*

FIGURE 2.37 *A simple type of thermosiphon solar collector uses the solar collector as a heat store. A sealed, well-insulated box, of which the top surface is the absorber, forms a solar heating system without mechanical parts (control and circulation), which does not need an electricity supply. The box contains water used as the heat store, and fresh water is led in a closed pipe (heat exchanger) through the store. Installation is simple and operation reliable. A suitable position for the solar collector is in direct connection with the place where it will be used*

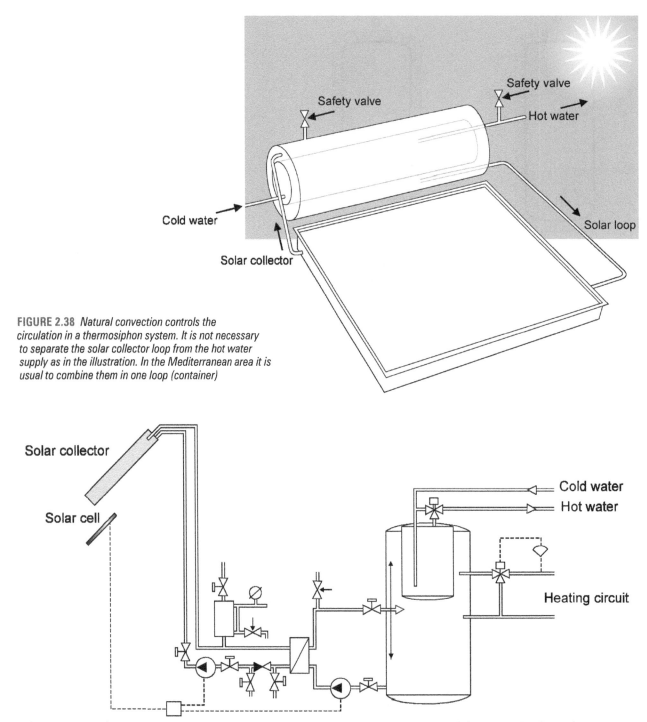

FIGURE 2.38 *Natural convection controls the circulation in a thermosiphon system. It is not necessary to separate the solar collector loop from the hot water supply as in the illustration. In the Mediterranean area it is usual to combine them in one loop (container)*

FIGURE 2.39 *Low-flow system with an external counterflow heat exchanger controlled by a solar cell. The solar cell varies the flow, which means a faster flow with more intensive solar insolation, and the system is independent of electricity supplied from the grid. A separate control unit is not necessary*

it can be necessary to increase the flow at high temperatures), and it is important that the storage tank really is temperature stratified, which makes higher demands on the installation work. Even if low-flow systems are still in the development phase, the technology is here to stay. At present, low-flow technology is used primarily in large-scale solar heating applications, such as the district heating plant in Falkenberg (see Figure 3.1).

A new technique is being developed for storage tanks in which all the heat transfer (both solar heat and DHW) takes place in external counterflow heat exchangers. In this way maximum temperature stratification is obtained and the function of the tank is optimized. The temperature conditions are extremely good for solar heating, and heat losses are minimized. As production costs go down this will become an interesting system design for storage tanks in general and 'solar stores' in particular.

2.7.4 DRAINBACK SYSTEMS

Drainback systems (Figure 2.35 (C)) are a simple way of avoiding risks of freezing and boiling in the solar loop; also, normal water can be used as the heat transfer fluid. One of the first drainback solar heating systems for the single-family house market was based on a technique using a double-jacketed hot water tank as a heat store. The solar loop's heat transfer fluid circulated in the outer jacket, and there was fresh water in the inner jacket (copper lined or enamelled). The drainage vessel used to collect the heat transfer fluid when the system was not in operation was placed in a frost free space, directly over the storage tank or in an attic space.

A drainback solar loop works in exactly the same way as a pressurized pumped solar heating system. When the heat in the solar collector exceeds the reference temperature in the heat store the pump starts. The heat transfer fluid is raised from the drainage vessel up to the solar collectors. When the temperature is evened out or when there is a risk of too high a temperature the pump stops and the heat transfer fluid runs back down to the drainage vessel. It is important that the pipes slope down throughout their length. The air space in the drainage vessel is used as expansion space. Always follow the supplier's instructions and the rules and regulations in force in the relevant country.

Considerations for drainback systems are as follows:

■ The drainage vessel must be positioned so that it is free from frost, and lower than the solar collectors and the parts of the pipe runs that can be exposed to frost.

■ The pump in a drainback system must be suitable for the height rise. The capacity of the pump must be sufficient to lift a head of water from a state of rest in the drainage vessel to the top of the solar collector. The pump must be constructed so that the heat transfer fluid can run back when the pump is turned off.

■ A drainback solar collector must have an absorber construction with a common lowest point; this also applies to the parts of the solar loop that are to be drained. The solar collector's absorber and the whole of the pipe circuit must be completely corrosion-proof (of plastic, copper or acid-resistant stainless steel).

■ Just as in a pressurized system, a drainback system should have a safety valve. Even if there is air in the system, steam can be formed (the fluid boils), which means that the system must be safeguarded against high pressure.

◼ 3. **Solar heating applications**

In spite of a comparatively small home market, Sweden is at the forefront of the field of solar heating, even though conditions have not been particularly favourable. A difficult competitive situation (the energy price) and a stop–go energy policy are two of the obstacles that have hindered the use of solar heating in Sweden.

Nevertheless, Sweden can boast some of the largest solar collector fields in the world (Table 11.1). The large-scale solar heating plants that are being built in Europe are based largely on Swedish technology and Swedish know-how. The knowledge gained from these plants has stimulated the development of other solar heating systems, for example for the single-family housing market.

Swedish companies and researchers have been working on an integrated solar collector model for multi-family dwellings for many years. This has developed into a solar collector that is mounted directly on the roof, which has many advantages. Both design and construction work are minimized, so that costs can be kept down. Purchase is simpler, and the demarcation between the installation work, the operational responsibility and the guarantee conditions is simpler and clearer.

A normal roof on a single-family house in Sweden receives five times as much energy as the house uses during a year. The sun's radiation impinging on the earth in 5.5 minutes can be said to correspond to the accumulated energy supply (consumption) of the world for a whole year (100,000 TWh in 1995).

Now there is a solar heating technology that is suitable for conventional heating systems. There is the know-how for designing good systems, and information on the energy supply from solar collectors is ensured, for example through SP – the Swedish National Testing and Research Station in Borås – and other testing stations throughout the world. It is important that the natural use of solar energy should be emphasized, and that it is not only found within the housing sector. Other uses can be for camping sites, sports facilities and similar seasonal users. That solar heating is not utilized to a greater extent and in projects where it seems obvious is, in many respects, a problem of information.

As with all (new) technology it is important to be involved in the design at an early stage. Solar collectors need roofs or ground areas orientated to the south and free from overshadowing. Space is required for storage tanks, and pipes have to be laid between the solar collector and the system equipment. To optimize profitability and operation it is important for solar heating to be included in the heating system at the design stage.

3.1 **Multi-family dwellings**

Solar heating systems for multi-family dwellings are based on conventional building services technology. The basic installation consists of solar collectors that heat a water storage tank, but these are combined with a boiler for biofuel, oil or gas. All charging and discharging of

FIGURE 3.1 *Aerial photo of solar collector field, Falkenberg Municipality, on the west coast of Sweden. Total solar collector area 5500 m²*
Photo: Lars Andrén

heat for space heating and domestic hot water uses the same storage tank, in the same way as in other solar heating systems. Since the first systems were installed at the beginning of the 1980s, over 10,000 m² of solar collectors have been installed on multi-family dwellings throughout Sweden.

Integrated solar collectors are the most interesting development in current solar heating systems for multi-family dwellings. They are either site built or prefabricated, and often cover the whole of the south-facing roof surface. They work both as a roofing material and as a producer of heat. One of the latest solar collector types is prefabricated by a house construction company. The solar collector module is integrated into the roof construction and is delivered to the building site as a complete element.

Unfortunately there have been far too few applications of roof-integrated solar collector technology for multi-family housing during the 1980s and 1990s. In Kungsbacka municipality, however,

EKSTA Bostads AB (the housing company) has consistently used solar heating in its new buildings, right from the beginning of the 1980s (Figure 3.2). This has meant that EKSTA can now show lower heating costs per living area than similar companies elsewhere in the country.

3.2 Single-family dwellings

Solar energy technology is well developed for the single-family house market. System technology is on the way to being phased in as a natural part of the heating system. By utilizing a common storage tank, solar collectors today can be a supplement to any type of energy (heat source).

During the years 1991–96 the installed solar collector area increased by about 20% per year in Sweden. Within the EU there was also a steep increase at the beginning of the 1990s. The sales of glazed solar

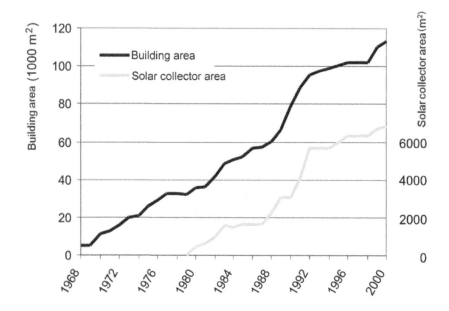

FIGURE 3.2 *EKSTA Bostads AB (housing company) in Kungsbacka has consistently used solar heating since 1980. Today, EKSTA has over 6000 m² of solar collectors in operation. It is interesting that EKSTA has continually invested in solar heating in new buildings and thereby learnt the technique, which has led to profitability in the projects Source: Jan-Olof Dalenbäck, Chalmers University of Technology, Göteborg, Sweden*

collectors increased from slightly under 500,000 m² in 1990 to slightly over 1,000,000 m² in 2000 (see Figure 11.3). However, sales stagnated towards the end of the decade. The greatest expansion is taking place within the single-family house market, where the DIY movement accounts for an important part of the annual sales. The number of brands is increasing, and installation engineers are becoming more familiar with the technology. Expansion can be expected to continue if nothing unforeseen occurs.

There are many ways of using solar heating systems in single-family houses – anything from simple shower installations to sophisticated solar heating systems for both domestic hot water (DHW) and space heating (Figures 3.3–3.5).

Generally, efficiency and profitability are best in combined systems, in which the production of DHW and the distribution of space heating is based on the same storage tank. The solar heat is also connected to another heat source, for example from electricity, wood or oil; gas or coke could also be used. The heating system is flexible, and the most appropriate technology can be used for each type of energy, with a full range of

FIGURE 3.3 *Ordinary DHW system with solar collector connected to a hot water tank. The solar collectors cover slightly over 50% of the annual DHW demand for a normal Swedish household of four persons Photo: Lars Andrén*

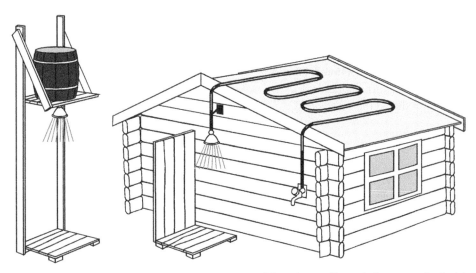

FIGURE 3.4 *The simplest ways of using solar heat are often unsophisticated, ingenious and home-built systems. On the left, a barrel (or container of some sort), which serves both as absorber (solar recipient) and heat store. The water in the barrel is heated up by the sun to be used as shower water. On the right, a solar heating system consisting of a black plastic hosepipe laid on the roof of a summer cottage. Incoming cold water is led through the pipe to raise the shower temperature*

combinations on both the production and distribution sides.

During the summer months solar heating can supply up to 90% of the DHW demand in most of northern Europe (where the average temperature is high enough). During the winter period the coverage is at best 20%. A properly functioning solar heating system can cover 50–60% of a normal family's (in Sweden – four people) annual DHW demand. In a combined system, solar heating can achieve an annual coverage of 20–25% of the total demand for heating and DHW. In a new single-family house, in which only a small part of the demand is for space heating, the solar fraction increases, but it is lower in an older house with a comparatively larger demand for space heating. It is important to take the heat load for space heating and DHW into account in the design. It is also important to pay attention to the risk of the growth of legionella bacteria, by always ensuring the prescribed minimum temperature, not just in the water store but right out to the draw-off point.

The efficiency of solar collectors has more than doubled and their cost has halved since the early 1980s. The systems are more efficient, with a higher

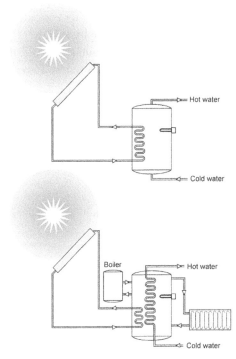

FIGURE 3.5 *Two types of system. Top: DHW system in which the solar heat is stored directly in a hot water tank. Bottom: combisystem, in which the solar heat can be used for both space heating and DHW.*

performance and longer life than those of just a few years ago. The system technology is suited to conventional heating, ventilation and sanitation technology. Operating packages (see Section 2.4.2 and Figure 2.21) have simplified the installation work and made it safer, and this has reduced the costs.

3.3 Outdoor pools

The heating of pool water is one of the most common and most profitable areas of use for solar collectors in the world. In California (USA) there is even a law forbidding forms of heating other than solar heating for outdoor pools.

In colder Northern European countries, the outdoor bathing season coincides with the solar season. The working temperature of a pool plant gives the best possible efficiency for a solar heating system. The possibility of using low-temperature heat allows simpler types of solar collector to be used, which makes for extremely favourable economics, in terms of payback time.

An outdoor swimming pool without auxiliary heating will seldom be warmer than 20°C in Sweden. To reach a desired pool temperature of around 25°C takes about 0.5 TWh heat per year in Sweden. Auxiliary heat production is normally by electricity or oil.

There are two types of solar heating system for swimming pools. The first uses unglazed solar collectors. The pool water is used as the heat transfer fluid in the solar collectors, and delivers heat direct to the pool without a heat exchanger or interim storage.

The alternative is to use high-temperature (glazed) flat-plate solar collectors in an indirect solar heating system (for indoor swimming baths and combined indoor and outdoor baths, for example). In this type of system the heat production from the solar loop is transferred via a heat exchanger to heat both the pool and shower water (Figure 3.6). This system solution is most suitable where there is a winter load in connection with the outdoor pool, such as an adjacent sports hall.

3.4 Large-scale solar heating technology

Large-scale solar heating plants have been developed and built in Sweden since the beginning of the 1980s: the latest was in Kungälv in 2000.

FIGURE 3.6 *Vislanda swimming baths, Alvesta Municipality, Sweden: 200 m² of high-temperature solar collectors heat the pool and shower water and replace, first of all, auxiliary heat from oil-fired boilers*
Photo: Lars Andrén

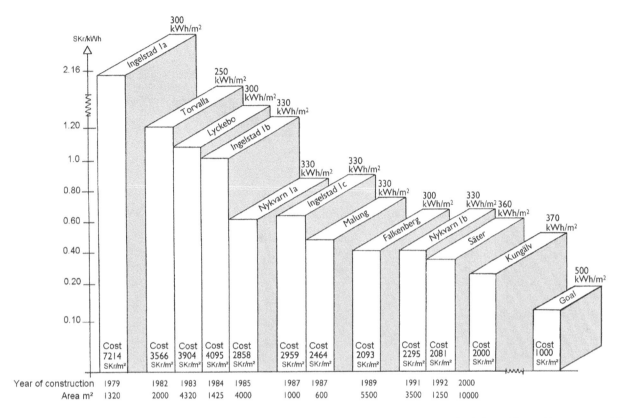

FIGURE 3.7 *Costs of large-scale solar heating projects in Sweden. The cost is based on a 20-year payback time with a real interest rate of 6%, in 1994 prices, except for the Kungälv project, which is given in 2000 prices (9 Skr = approximately €1). The cost of the heat store is not included. Ingelstad 1a was a field with concentrated solar collectors; this technique is no longer used in Sweden. Lyckebo, Ingelstad 1b and Nykvarn 1a consist of flat-plate solar collectors with a double layer of Teflon (convection barrier). The solar collectors in Ingelstad 1c, Falkenberg and Nykvarn 1b are constructed with a single layer of Teflon. In Säter, the solar collectors are of site-built large module construction. The Swedish Council for Building Research (BFR, now Formas) has taken part in all the above projects except the Kungälv project, as part of the Swedish research work. Source: Jan-Olof Dalenbäck, Chalmers University of Technology, Göteborg, Sweden*

Such large-scale projects have made it possible to measure, analyse and research all the components. Today, projects with short-term storage for a heating plant with an annual consumption of about 60 GWh are considered as being the nearest to profitability. The best prospects are for connection to oil-fired heating plants. Seasonal storage systems, as a supplement to larger oil-fired heating plants, are also of interest. A reduction of costs or an improvement in performance by 20–25% is required to reach a commercial breakthrough in Sweden.

3.5 Camp sites

There are great opportunities for solar heating to meet much of the DHW demand on camp sites (Figure 3.9). However, it is always important to equip the system with a temperature guarantee in the form of another heat source, such as a solid-fuel boiler or electric heater.

To achieve a system solution with optimal operation and coverage, basic information is needed about occupancy (number of places and persons during the high season), number of showers (draw-off points that

FIGURE 3.8 *1025 m² of Swedish-manufactured flat-plate solar collectors connected to a district heating plant in Saltum, Denmark. The solar collectors are mounted on a concrete footing, which is placed on the ground without any preparatory ground work in order to minimize investment costs*
Photo: Lars Andrén

demand hot water), maximum discharge (estimate of the extent to which all the draw-off points will be used at the same time), usable roof areas (free from overshadowing), space for storage tank, total heat transfer rate demand, and preferred type of auxiliary heating.

3.6 Sports facilities

Solar heating for sports facilities (Figure 3.10) is based on the same conditions and system solutions as for camp sites (see checklist in Box 3.1). The differences are that sports facilities are often used for a greater part of the year and that maximum discharge (the heat demand) is more intensive, which makes higher demands on the auxiliary heating.

Sizing is based on the following basic factors:

■ occupancy (number of persons)
■ number of showers (draw-off points demanding hot water)

Box 3.1 Checklist for camping and sports facilities

Unshadowed roof areas	yes/no	Estimated solar collector area	m²
Space for storage tank	yes/no	Storage tank volume	m³
No. camping places	no.	DHW production	kWh
No. showers	no.	Maximum draw-off	kWh
Maximum draw-off in hours	hours	Auxiliary heat requirement	kW
Estimated DHW demand	kWh/24 hrs		
Auxiliary heat (check)	☐ wood ☐ pellets	☐ gas ☐ oil	☐ electricity ☐ ....

FIGURE 3.9 *Skärshult camp site, Hyltebruk Municipality, Sweden: 10 m² of solar collectors*
Photo: Lars Andrén

- maximum discharge (time estimate of the extent to which all the draw-off points will be used at the same time)
- usable roof areas (free from overshadowing)
- location of storage tank
- total heat transfer rate
- preferred auxiliary heating.

It is important to consider the external conditions when a solar collector is to supplement the heat supply in a sports facility. In some cases the sports activities may not be as intensive during the summer as during the winter period, or vice versa. Furthermore, it is of great importance to make a careful study of the heating equipment that the solar heating is to be based on.

Sports facilities that house 'summer' sports such as football and athletics are of special interest. In some of these sports facilities the layout is also suitable for solar heating, with relatively small, individual units placed directly adjacent to the draw-off points. In this way the connection of the solar collectors is both simple and cost-effective.

3.7 Schools

If there are no summer activities in the school buildings then there is not much point in installing solar heating. Schools in Sweden are empty of pupils and have no activities for almost 10 weeks during the sunniest period of the year.

FIGURE 3.10 *Solar collectors of 35 m² area for DHW for a sports facility in Hyltebruk Municipality, Sweden*
Photo: Lars Andrén

However, a solar heating installation may fulfil an educational function and be motivated for this reason. The solar heating system gives both practical and theoretical knowledge. The measurement and computer logging of measured values can be of interest, as well as the practical installation work.

In schools with summer activities the situation is different, of course. Sizing is based on the consumption during the summer and the load, primarily for DHW, that can occur. In some cases, as described later in Chapter 6, heating an adjacent outdoor pool can be of interest.

Otherwise it is normal for solar heating to be integrated into an existing heating system in the school.

This means finding the best possible way of connecting the solar heat, for example the optimal installation of the solar collectors in the system and whether any of the existing system components should be replaced. When the investment is made, the older system configuration can be renewed at the same time.

3.8 Industry

Companies concerned in environmental work can have a special interest in solar heating, but economy is, of course, decisive for the will to invest. This is the main reason why solar heating is of less interest to the industrial sector: the payback time is too long.

The food industry can be an exception, as it has a constant need for hot water. The ice-cream industry, however, has peak production during the summer months, which means a large consumption of hot water then. At the same time it can be an advantage for a freezing plant to produce hot water with the surplus heat (for example via a heat pump) gained in the refrigerating process.

There are many interesting industrial projects in Europe. For example, in the Netherlands there are two companies where solar heating is used to dry tulip bulbs, and another where food cans are washed out with solar-heated water.

■ 4. Solar heating for multi-family dwellings

4.1 Use

In many countries, solar energy technology for multi-family dwellings has developed in parallel with solar heating for single-family houses. Basically, solar heating systems for multi-family and single-family dwellings are built up of the same component parts and system technology. Sizing and key ratios are not completely different from one another either.

One of the most interesting solar heating applications is the integration of solar collectors in the building envelope when this is to be renewed anyway, for example when altering flat roofs or wall claddings. Both water-based and air-based solar heating systems can be of interest. In the latter case the solar heating is normally used for ventilation air or space heating, preferably near the position of the fresh air intake to the building.

A decisive factor in whether or not solar heating is of interest in refurbishing is whether the building envelope needs attention for reasons of heating technology or maintenance, or for architectural reasons.

The greatest opportunities for optimal installation and operation are in new building. It is important to check the options and try to find an advantageous position for the solar collectors, space for the storage tank and other heating equipment, suitable pipe runs, good combination of energy types, and good economy.

When solar collectors are to supplement an existing heating system, further demands are made on the survey work. When the roof covering is to be changed it can be natural to choose solar collectors as a roofing material, which is particularly interesting if flat felt roofs are to be altered to ridged roofs. Here there are often good conditions for solar collectors, from both the financial and the technical points of view.

Start with the storage tank, regardless of whether it is a question of a new heating plant or supplementing an existing system. The best economic conditions are created if solar heating is used together with another heat source that requires some sort of storage tank.

Box 4.1 Input data: preliminary study

Existing buildings
- Collect available measured data.
- Total area of dwellings and other premises.
- Cold water consumption.
- Hot water consumption.
- Hot water circulation.
- Heat control in the summer.
- Existing equipment.
- Alternative positions for the solar collectors and other heating equipment.

New building
- Number of dwellings, usable floor area.
- Estimated DHW consumption and consumption for hot water circulation.
- Choice of system and auxiliary heating source.
- Heat demand and heat transfer rate demand.
- Choice of heating distribution and ventilation systems.
- Options for position of solar collectors and heating equipment, and space for pipe runs.

Thus the cost of the heat store does not have to be borne by the solar heating alone.

Box 4.1 lists the basic data that are required when making a preliminary study.

4.2 Roof-integrated solar collectors

It is advantageous for multi-family dwellings, in one or several blocks, with a common heating plant to use a solar heating system. It is important that the solar heating is included in the design at an early stage to optimize performance and minimize the extra cost of the installation. For example, the building orientation, the slope of the roof and the space for the heating equipment are of great importance.

The choice of the auxiliary heat source is important for determining the solar collector area. The design of the system is vital to the amount of heat that the solar collectors can contribute or replace. This, in its turn, gives the plant's coverage and profitability, factors that can seem to be independent but which are important

for the final result. When choosing a system it must be considered as an entity based on the prevailing conditions for each case.

A roof-integrated solar collector works both as roof covering and as heat producer. Southerly-facing roof areas free from overshadowing are necessary. Section 2.1, on orientation and output, shows how the production of heat differs from the optimal for different roof slopes and orientation.

At present there is an area of somewhat more than 8000 m^2 of solar collectors of the roof-integrated type in Sweden. Soon the installation of solar collector roofs will be a natural part of the building process. Integrated solar collectors will probably replace conventional roofing material where this is suitable.

4.3 Sizing

The solar collectors are sized primarily to give the heat required to cover the hot water demand during the summer months. When there is enough solar radiation

FIGURE 4.1 *Roof- integrated solar collectors in Furuby near Växjö, Sweden. Property owned by Värendshus AB*
Photo: Lars Andrén

during the spring and autumn, and the solar loop is connected to a storage tank, the solar collectors will also contribute to heating the dwelling.

To design and size a solar heating system, the dwelling's consumption of heat and domestic hot water and the distribution during the year must be known or be able to be estimated. The solar heating system is also affected by the accessible roof areas of suitable slope and orientation and space for the storage tank. See flowchart in Figure 4.2.

Normally solar collectors produce the most heat at high load, which means that an undersized solar collector plant gives a relatively good production of heat per m². There is normally no reason to oversize a solar heating system unless the aim is to turn off an auxiliary boiler for a certain period.

The normal solar fraction is between 15% and 25% of the total heat load (depending on the design of the system, the solar collector performance and the orientation, slope of the roof, length of piping and insulation). A good rule of thumb is that the solar heating should be designed to cover 40% of the domestic hot water (DHW) demand. Experience has shown that systems on this basis produce heat corresponding to 400 kWh/year per m² of solar collector area.

It is the summer load that directly determines the solar collector area. The volume of the storage tank is, in its turn, determined by the solar collector area and the choice of auxiliary heat.

4.4 Key ratios

The tables and information boxes below give, quickly and easily, an idea of the design of the system, suitable solar collector area and storage tank volume, and economy and energy. They can give a pointer to suitable solar heating systems. This can be a basis for continued design. All data given are based on Swedish conditions.

4.4.1 HEATING AND DHW DEMAND

There is great variation in the heating and DHW demand in new buildings. The normal annual energy demand is between 80 and 120 kWh per m² of dwelling

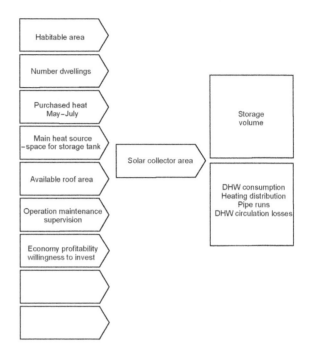

FIGURE 4.2 *Flowchart for sizing solar heating systems. Sizing of storage volume must be carried out in the right order and with a specification that fits the individual object*

area. The demand is greater in older buildings. The DHW share of the total heating load normally varies between 20% and 30%, depending partly on the inhabitants' pattern of consumption.

Table 4.1 gives guidelines for determining the DHW demand.

Table 4.1 Normal energy consumption for DHW in multi-family dwelling in Sweden

Flat size	Annual consumption per flat
1 bedroom, living room, kitchen (60 m²)	2000 kWh
2 bedrooms, living room, kitchen (80 m²)	2500 kWh
3 bedrooms, living room, kitchen (100 m²)	3300 kWh

4.4.2 SOLAR COLLECTOR AREA AND STORAGE VOLUME

The area of the solar collector can be calculated from the number of dwellings and the volume of the storage tank. When the solar collectors are intended only to produce DHW for a fairly small hot water tank (50 litres/m² solar collector), 4–5 m² of solar collector per dwelling is sufficient. However, it is normal for solar heat to be used in what is known as a *combisystem*: that is, for both DHW and space heating. In this case a solar collector area of 5–8 m² is often sufficient, but a larger storage volume often demands an increase in the solar collector area – otherwise the temperature rise in the storage tank would be too small.

The space needed for the components of the solar heating system (such as storage tank, heat exchanger, and pump) amounts to approximately 6–10 m² of floor area per 100 m² of solar collector area.

4.4.3 HEAT PRODUCTION

Heat production from solar collectors obviously varies with the make, the system (primarily how the storage tank works), the orientation and roof slope, and the number of sun hours during the year (for example, in Sweden there is generally more sunshine on the coast). A standard solar collector gives about 400 kWh/m² annually. The actual output varies, of course, with the performance of the solar collectors, and it is also affected by the design of the system in which they are used and by the type of heat source they replace. It is important to remember that the heat output is normally affected more by the system design than by the solar collector performance.

4.5 System combinations

The basis of the system design obviously varies with the individual conditions. Among other things, systems are designed differently depending on the size of the heat load. In general the volume of the heat store is determined principally by the solar collector area in relation to the actual summer load. A large solar collector and a large summer load necessitate a larger heat store, whereas a small collector and a small summer load need only a smaller heat store. The type of auxiliary heating and the annual and daily heating

demand also affect the choice of heat store and its volume, but to a lesser extent.

Solar heating can be used without a heat store, for example by connecting the solar loop directly to the district heating network. In this way no heat store is needed as the demand is often so large that the solar heat can be taken up directly, the volume of the district heating circuit being used as the heat recipient/heat store. The costs for the system will thus be lower, but with the disadvantage that the working temperature is higher, which gives lower efficiency and heat output.

A combination of solar heating and biofuel is common in Sweden (Figure 4.3). The storage tank is used both when burning wood and for the solar heating. The storage tank allows a longer combustion period each time the boiler is lit, thus increasing the efficiency and the life of the boiler, especially in smaller systems. Losses due to uncontrolled ventilation are smaller with longer combustion times, and combustion can also be at full power. As soon as the temperature in the solar collectors is higher than at the reference point in the storage tank the solar collectors can start to deliver heat. In this way the number of running hours for the solar collector increases, while the working temperature of the solar heating system can be kept low.

Solar heating can be combined with fossil fuels using a storage tank. Operation and efficiency can be improved, and the environmental impact from the fossil fuels is reduced. It is important for the storage volume to be suitable for the requirements of both the solar heating and the fossil fuel. The difference in the required volumes is less than with the combination of solar heating and wood-burning boilers.

The advantages of combining solar heating and other heat sources via a common storage tank are:

- a more flexible heating system
- increased independence
- predictable heat costs
- reduced vulnerability and greater flexibility (if the boiler breaks down the storage tank can be used as the heating unit)
- lower service costs for the boiler units

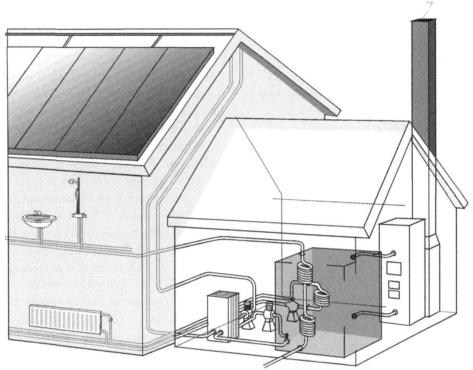

FIGURE 4.3 *Solar heating and solid fuel (wood) fired boilers are a good combination. Both the heat sources are combined with a common storage tank. It can be either pressurized or, as here, non-pressurized*

- longer life for the auxiliary heating unit
- better utilization and higher efficiency for the boiler units
- reduced environmental impact.

In larger systems the size of the storage tank is determined not by the requirement of the auxiliary heating but rather according to the conditions for solar heating. The function of the heat storage tank is to transfer heat during the day (over 24 hours), from a period of surplus to a period of demand. The extreme case is the system design in which all the solar heat produced can be used directly without a heat store.

4.6 Operation and maintenance

It is the duty of the manufacturers and suppliers to provide some sort of running and maintenance

instructions for the plant at the final inspection of an installed solar heating system. It is also important that the operation and maintenance manual includes a complete set of data sheets on all the components included in the system. Furthermore there should be instructions on tasks such as topping up the heat transfer fluid and setting operational thermostats and pumps, on the normal working pressure in the system, and on the recommended flow and temperature rise in the solar heating loop, as well as on safety regulations and warnings.

Box 4.2 gives a typical checklist for operators.

4.7 Summary

There is great interest in using solar heating in multi-family dwellings, and this interest is increasing. The

Box 4.2 Solar heating systems: checklist for operators

Daily supervision
▧ Check the operating and indicator lights on the control unit.

General inspection
▧ Pump operation. During longer periods of non-operation (during the winter, for example) test-run the pump.
▧ Temperature levels at the installed sensors, and in the control unit if this has a temperature display.
▧ Pressure in the solar loop. Check at the manometer on the expansion vessel.
▧ Read the heat meter, if there is one.
▧ Temperature rise in the tank during the day and also the temperatures when the heat exchanger is in operation.
▧ Read the meter showing time of operation for the pumps, running time for the solar collector system and the electric auxiliary heater if there is one.

Other inspections
▧ Carry out air bleeding of the system a few times per year if the pressure in the system has gone down.
▧ Check and clean the filter in the solar loop; at the same time clean the filter at the flowmeter, if fitted.
▧ Check the antifreeze in the liquid in the system once a year.
▧ Check the amount of inhibitor in the fluid in the system if recommended by the supplier.
▧ Make a visual inspection of the solar collector, looking for suspicious leaks. Check fixings and also the flashings between the solar collectors and the roofing material.
▧ Check the start and stop function of the solar heating control unit and also the fixing and insulation of the sensor.
▧ Check the insulation of the pipe runs between the solar collectors and the tank.
▧ Check the operation of the non-return valve (or alternatively the magnet valve).

conditions can be optimized in new building to give the best possible performance, low investment costs and thereby also good profitability. It is also a great advantage to integrate solar collectors when the building envelope, for example flat roofs or facade material, is to be replaced, or when it is necessary to replace the heating system.

A well proven technology is available for multi-family dwellings, in which the solar heat is combined with the auxiliary heat in a common storage tank and is thereby integrated into the whole heating system. Even if the solar collectors are spread out over several buildings the heat is collected in one central storage tank. It is important that solar heating is included at an early stage in the design work to be able to design it for the project, and to give guidelines for the orientation and slope of the roof, the space for the solar heating equipment (storage volume), and adaptation to the rest of the heating system. The solar collector area should be sized and designed to suit the rest of the heating system. The coverage that is to be achieved and the choice of the auxiliary heating source are determining factors. The solar fraction can be between 15% and 25% of the total heating load, depending on the climate zone and the desired goal. Normally the aim is a solar fraction of about 40% of the DHW demand, and this requires between 4 and 8 m^2 solar collectors per dwelling. For every m^2 of solar collector, 50–100 litres storage volume is generally recommended.

Solar heating can be combined with any kind of energy by means of a storage tank. The design starts by considering the demands made regarding, for example, running costs, maintenance or the environmental considerations. This, in turn, is a guideline for the choice of auxiliary heat. Before starting the design a survey must be carried out of the existing conditions and the owner's demands and wishes regarding the system. This will then form the basis of the system design.

There are various different types of solar collector that can be chosen for multi-family dwellings. The traditional flat-plate solar collector dominates the market for multi-family dwellings. In recent years a roof integrated site-built solar collector has been developed. This gives an advantageous price and is both a roofing material and a producer of heat. A further development is the manufacture of integrated solar collectors as roof elements. The solar collector is delivered as a complete roof element, which is then mounted directly on the roof trusses.

■ 5. **Solar heating for single-family dwellings**

5.1 **Use**

There are several areas of use for solar heating in single-family houses. The most common is called a *combisystem* for space heating and DHW: here the solar heat is combined with other types of energy in a storage tank (Figure 5.1(A)). Solar heat can also be used just for DHW (Figure 5.1(C)). In houses with electric resistance heating, this is the only way of utilizing solar heating. There is also a possible market for solar heating for outdoor pools for single-family houses (Figure 5.1(B)).

With new building, or when changing the heating system, the system can be adapted to solar heating without causing extra costs, or the conditions for the rest of the system deteriorating. It will be both financially and technically advantageous to start from a storage tank and connect an optional source of energy in combination with solar heat (Figure 5.1(D)).

The combination of solid fuel – wood, briquettes or pellets – and solar heating is interesting for many reasons. All combustion is least effective during periods of low load: that is, during the summer when solar radiation is greatest. With solar collectors, the period for which the auxiliary heating system is used can be almost halved (in southern Sweden, around latitude 56° N). A storage tank allows longer running times giving better conditions for combustion and improving the efficiency of the system. This also reduces the total environmental impact.

It is particularly interesting to add solar collectors to single-family houses with existing storage tanks. Solar heat can be connected relatively easily to the storage tank, either by an external flat-plate heat exchanger or through a heating coil in the tank. The solar collector area is determined from the volume of the storage tank. With larger tank volumes (over 1500 litres) attention must be paid to the operation of the system and the suitability of the solar collector area for the volume of the storage tank. One alternative is to make several connections to the tank at different levels.

5.2 **Sizing**

The household's summer load, and in particular the DHW demand, determines the design and size of the solar heating system. Other factors that affect the choice of system and the solar collector area are the volume of the storage tank, the available roof area, other heating equipment (system), and possible extra consumption in the summer (such as an outdoor pool, a guest cottage with a shower, or heating in the basement to keep it free from damp). With a good-size solar collector area the storage tank is heated more quickly, which means that auxiliary heat can be turned off earlier.

When the size of the solar heating system is determined according to DHW consumption this is based on a normal Swedish family (four persons) using slightly less than 5000 kWh per year (10–14 kWh/24 hrs) for the production of DHW.

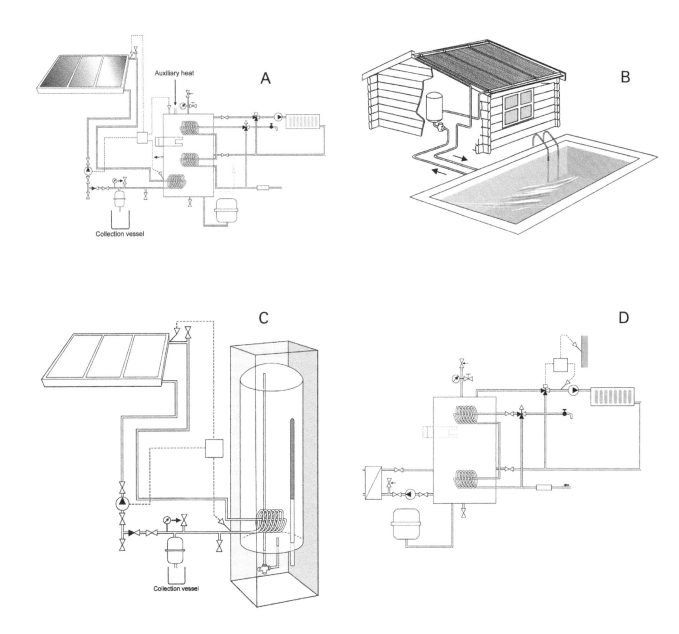

FIGURE 5.1 *Four system designs. (A) The solar collectors are connected to a combisystem in which the solar heat is used for both space heating and DHW. Optional auxiliary heating can be connected to the storage tank. (B) An outdoor pool is heated by unglazed (low-temperature) solar collectors. The heat transfer fluid (the chlorinated pool water) passes directly through the solar collectors. A simple type of system, which is very good value for money. (C) An ordinary DHW system with 4–6 m² of solar collectors connected to a hot water tank of 250–300 litres. Solar heating covers at least 50% of the annual demand. (D) The heat from the solar loop is transferred to the storage water. This is a normal way of transferring heat when solar heating is connected to an existing storage tank. In the same way a flat-plate heat exchanger can be used to transfer the heat directly from the solar loop or from the storage tank to a pool circuit. It is important to choose the correct type of heat exchanger, suitable for the heat transfer fluid and the working temperatures that occur. Always follow the supplier's instructions and the rules and regulations in force in the country*

The 'normal' solar collector – that is, the flat-plate solar collector – produces approximately 400 kWh per m² of solar collector per year in a well-functioning system. Thus with a solar collector area of 4–6 m², solar heating produces half of the yearly demand for DHW. As the DHW demand is relatively constant during the year this is a suitable basis for calculating the size.

In its turn, the volume of the storage tank is based on the solar collector area. Normally this is based on every m² of solar collector requiring 50–100 litres of storage volume. A normal DHW system with 5 m² of solar collectors requires a volume of 250–300 litres. In a combisystem, which is often a better system solution, a solar collector area of between 8 and 12 m² and a storage tank of 500–750 litres are normal.

In some cases, the auxiliary heating (wood burning) requires a larger storage volume. It can then be necessary to increase the storage volume with a further storage tank. In this way the system has a master tank (with technical equipment) and a slave tank (parallel coupled) (Figure 5.2). The master tank contains the coils for DHW and the outlet for the space-heating circuit. The slave tank is used only to increase the volume. The advantage of dividing the storage volume into two units is that the solar heat can operate with an optimal volume at the same time as the volume requirement of the wood-fired boiler can be met. Furthermore, heat losses during the summer months are minimized when only the master tank is used. The main heat store can also be placed inside a larger store. However, this requires careful sizing and makes great demands on the control of the solar heating.

The average energy consumption to cover space heating and DHW demand in a single-family house in Sweden is around 20,000 kWh/year. Newer single-family houses can have a lower annual consumption (about 15,000 kWh) (Figure 5.3). The consumption of older houses can be considerably higher, depending on such factors as size, standard of windows, insulation, and geographical position (Figure 5.4).

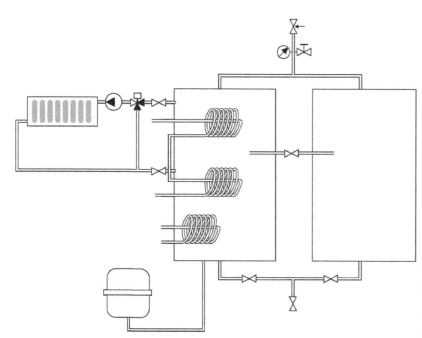

FIGURE 5.2 *The master tank is the part of the storage system that receives all the heat produced and then distributes it in the form of DHW and space heating. The slave tank is used only to increase the volume. By coupling both the units with large-diameter pipes (25mm), fitted with shut-off valves, optional storage volume (adjusted to the season: 2/4, 3/4 or 4/4) can be chosen. Stratification in both the tanks is by natural convection, and no mechanical transfer of heat is necessary*

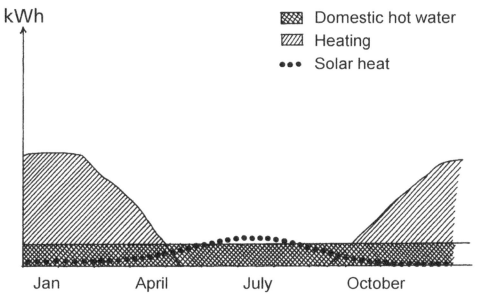

FIGURE 5.3 *The heat demand in a fairly new, single-family house is very suitable for solar heating. The solar fraction can be as much as 30% in total. Curves shown here for: 120 m² living area plus basement; 10,000 kWh for heating; 5000 kWh for DHW; 5 m² solar collectors; 300 litre storage tank volume; 50% solar fraction for DHW*

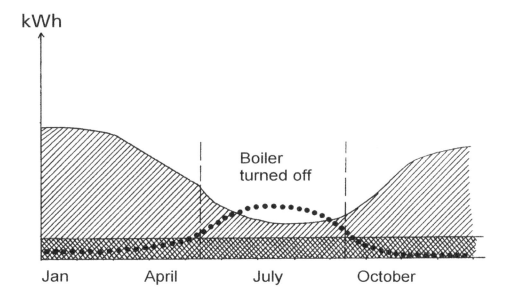

FIGURE 5.4 *In an older single-family house with a larger heat demand the solar fraction is smaller. Shown here: 120 m² living area plus basement; 15,000 kWh for heating; 5000 kWh for DHW; 8–15 m² solar collectors; 500–1500 litre storage tank volume; up to 20% solar fraction in total. An oil-fired boiler with low efficiency gives favourable conditions for solar heating, as the oil-fired boiler can be turned off for 4–6 months. During the summer the solar collectors can give excess heat, which can be used, for example, to keep a basement free from damp. Solar collectors together with an electric heater halve the period when the boiler must be in use. Solar heating alone can save up to 1 m³ of oil per year*

Table 5.1 Example of sizing of different heating systems for colder climates in Northern Europe

System	Storage volume (litres)	Area solar collector (m²)	Heat production (kWh/year)	Dwelling type	Solar fraction (DHW + space heating demand) (%)
Wood-fired boiler	2 × 750	12	4800	Older	20
Electricity	500	8	3200	1970s	29
Oil/gas	500	8	3200*	1950s	18

*Without taking improved efficiency of the auxiliary heat source into account. The heat reduction for the auxiliary heat source will probably be much larger and the solar fraction thus larger.

The storage volume must be sized according to the rated output of the wood-fired boiler: that is, the volume of the combustion chamber or possible power output/heat production.

5.3 Key ratios

Key ratios are intended to give a basis for quick design and sizing. Tables 5.1 and 5.2 show the energy consumption for different types of dwelling and the way the solar fraction is affected by the total consumption, based on the same heat production from the solar collectors. It is not certain that the output from the solar collectors will be the same in all three cases. The design of the system and the consumption pattern can affect the solar output very considerably.

Using Tables 5.1 and 5.2 it is possible to obtain suitable solar collector area and storage volume (for a hot water tank or storage tank) based on different requirements and conditions. There are variations in solar collector areas and storage tanks for different makes. Think of the tables as generally applicable and a good basis for continued design. All the data are based on Swedish conditions.

5.4 System combinations

There are various factors to be taken into account when choosing a system. Every consumer is unique. There are different requirements and wishes regarding investment costs, comfort, running costs, opportunities for service, environmental impact, accessibility and many other things. The design of the heating system requires a comprehensive view, in which the various wishes and factors must correspond.

Table 5.2 Typical energy demand in Swedish single-family houses

Energy consumption	Normal house	Newly built house	Older/larger house
Household electricity	5000 kWh	4000 kWh	5000 kWh
DHW	5000 kWh	5000 kWh	5000 kWh
Space heating	10,000 kWh	6000 kWh	20,000 kWh
Total	20,000 kWh	15,000 kWh	30,000 kWh
Solar heating 10 m²	4000 kWh	4000 kWh	4000 kWh
Solar fraction	20%	27%	13%

5.4.1 SOLAR HEATING AND ELECTRICITY, STORAGE BOILER

Combining solar and electric heating is of interest with a flat-rate tariff. If differentiated electricity tariffs are available, solar heating is in a difficult competitive situation as the summer tariff is normally set relatively low. Furthermore there is a temperature conflict during the period November to March, when cheap night-time electricity is to be stored for consumption during the day, which requires high temperatures (80–95°C). There is also a conflict in the size of the store, where the off-peak electricity requires larger volumes and the solar heating relatively small volumes. When sizing the storage volume according to electricity on a time tariff (that is, variable electricity prices) the following should determine the choice of volume:

■ power demand of the property
■ accessible power (main fuse)
■ number of hours off-peak rate (night-time)
■ available space for storage tank.

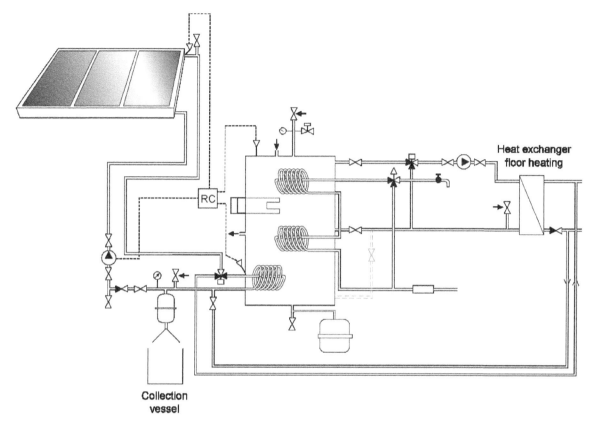

FIGURE 5.5 *Combined solar heating and floor heating. There are a number of advantages in transferring the solar heat directly to the floor heating system. The coverage increases, and the solar heating can be used for space heating for a larger part of the year than in conventional systems. To work well the floor heating system should be professionally designed. It must be quick reacting, and the control equipment must fulfil special requirements*

When combining electrical heating with solar collectors a smaller storage volume (500–750 litres) is an advantage. The tank works as an electric boiler, from which all heat and DHW is supplied (Figure 5.6).

The power demand of the property determines the size of the electric heater. When the property has a small power demand (6–9 kW) it is sufficient to place an immersion heater in the upper part of the storage tank, immediately under the upper DHW coil. When the power demand is greater than 9 kW it can be an advantage to use two electric heaters, one in the upper part of the tank and one in the lower part (just above the lower DHW coil). The heater in the upper part of the tank guarantees the temperature, and can be in use

throughout the year. The heater in the lower part of the tank guarantees the power, and is used only during the winter period. It is important that it be turned off during the summer as it spoils the temperature stratification in the tank.

Compared with a conventional electric boiler the investment costs for a storage boiler are lower and the installation is less vulnerable because the component parts are normally more easily accessible. This also makes service and maintenance easier. The use of a storage tank as the base for the heating system also reduces the vulnerability of the system. When a storage tank is used it is considerably easier to change energy sources. Thus the system is not restricted in the same

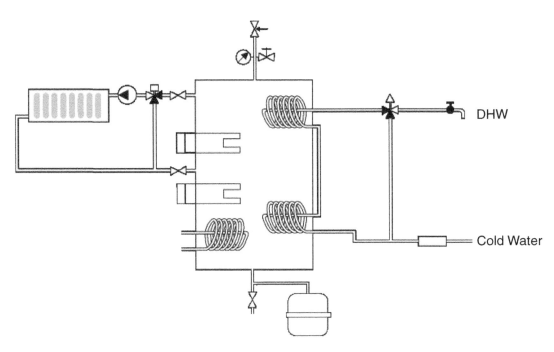

FIGURE 5.6 *When a storage tank is to be used as an electric boiler with single-tariff electricity it is sufficient to place an electric heater in the top of the tank, on condition that the power corresponds to the total power demand of the house for space heating and DHW. If the tank is to be used for differentiated-tariff electricity, both the volume and the power must be suited to this. A way of increasing the storage volume in the tank in the winter is to place an extra electric heater in the lower part of the tank. Always follow the supplier's instructions and the rules and regulations in force in the country*

way as when a single type of energy is chosen for the heat supply. The volume of the tank allows for the connection of other types of energy in the future.

5.4.2 SOLAR HEATING AND FOSSIL FUELS (GAS/OIL)

The storage tank allows longer running times for gas- or oil-fired boilers. A tank volume of about 500 litres is recommended for ordinary single-family house installations, but it must of course be suitable for the actual solar collector area and the operation of the boiler (to attain the longest possible running times). The storage tank has the advantage that it allows electricity and oil to be chosen according to the price level and availability (see Figure 5.7). It is important for the storage tank to have a core function as distributor of heating and DHW when it is connected to an existing oil-fired boiler. Investment in a storage tank is also important as a safety measure. An older, oil-fired boiler

can break down when least expected. During a transition period a storage tank with an electric heater can ensure the provision of space heating and DHW. Besides this, the heating system's flexibility and total efficiency are increased (heat losses are reduced and the boiler's running time is increased). There are also special combined systems for gas and solar heating (see Figure 5.8).

5.4.3 SOLAR HEATING AND PELLETS

Pellets are a modern form of biofuel. During the 1990s interest in burning pellets increased because of the cost advantage of the fuel, its convenience, and the great environmental advantages.

There are two main alternatives for heating with pellets. A pellet stove can be used to heat the house, as an individual heat source such as a closed stove or tiled stove, almost always in combination with other forms of

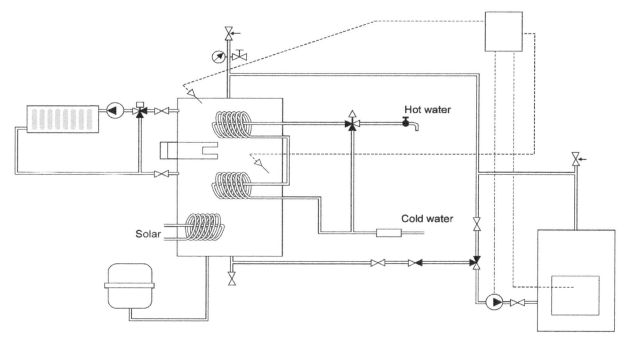

FIGURE 5.7 *It is an advantage if the new storage tank is fitted with controls for space heating and a DHW unit even if the existing oil-fired boiler has these. This is partly because it can be used as a standby to reduce vulnerability and partly because it is simpler and cheaper to change the boiler when necessary. Efficiency and environmental impact are other factors that are influenced in the right way when the oil-fired boiler can charge a storage tank. During the time when the efficiency of oil-fired boilers is at its worst, solar and electric heating cover space heating and DHW*

heating. The alternative is a pellet burner, a piece of equipment that is connected to an existing or a new boiler. The pellets are burned in the pellet burner, and the heat is taken up in the boiler unit, which then supplies space heating (often in a water-based system) and DHW.

Another, less common, alternative is to alter an existing wood-fired boiler. The wood-fired boiler must fulfil the technical demands made by pellet combustion.

A heating system with a pellet burner connected to the boiler is, in many ways, similar to an oil-fired system. The pellets are fed directly from the store to the burner; the heat is produced in the boiler unit and can be stored in a small storage tank. The system can be designed according to individual requirements (such as size of store and transport length of the feed screw).

From the point of view of solar heating, combination with a pellet burner is the most interesting (see Figure 5.9). The advantages of solar heating in summer can be combined with the biofuel's environmental, cost and power advantages. Solar heating can be connected in several different ways, depending on the user's consumption pattern and the size of the pellet burner. It is most common to attach the pellet burner directly to an existing oil-fired boiler. A storage tank allows improvement of the efficiency of pellet combustion, while the capacity of the pellet burner can be minimized. A storage tank with an electric heater is very advantageous for the system. Pellet combustion can be completely turned off during the summer, and at the same time the solar collectors have been added to the system in the best possible way.

5.4.4 SOLAR HEATING AND WOOD

A wood-burning boiler with large heat capacity (>20 kW) requires a large heat store to store the heat

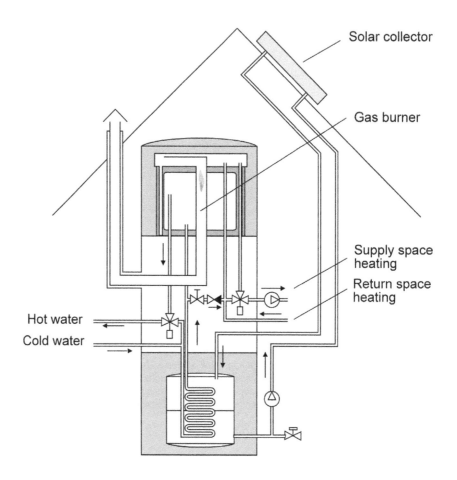

FIGURE 5.8 *On the European continent there are special combination systems for gas and solar heating. The solar and gas heating together cater for space heating and DHW production in small separate units. There are a number of different systems available. The illustration shows the construction of one such system*

Solar collector

Gas burner

Supply space heating

Return space heating

Hot water

Cold water

over the day from the time of production to the time of consumption. The recommended storage volume is determined primarily by the wood-fired boiler's heat capacity (the volume of the combustion chamber) and by the power demand of the property and the accessible space for the storage tank.

The considerably smaller volume required for solar heating is solved by the store's being divided up into several units: a master tank (with technical equipment) of about 500 litres and one or several slave tanks (Figure 5.10). The volume of the slave tank or tanks is determined according to requirements, and it is/they are coupled in parallel with the master tank. In this way the storage volume can be adapted at different times and according to the type of energy that is being used (or according to the season).

There are great advantages in supplementing a wood-burning system with solar heating. The solar collectors will account for most of the heat production during the summer period, when the system efficiency of the wood-fired boiler is at its worst. If the wood-fired boiler does not have to be lit in the summer, the system will be more convenient (less attention and less wood chopping), more efficient and, above all, more environmentally friendly.

5.4.5 SOLAR HEATING AND SEPARATE HEATING UNITS

A strong argument for an individual heating unit is that today's well-insulated single-family houses do not have a large enough heat load for an ordinary wood-fired boiler. New single-family houses generally lack the space for both a boiler and a storage tank.

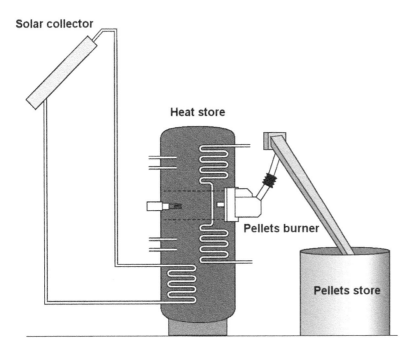

Solar collector

Heat store

Pellets burner

Pellets store

FIGURE 5.9 *New design of storage boiler with solar and pellets as the basic heat sources and an electric heater as reserve. The boiler is based on a traditional cylindrical storage tank with built-in coils for solar heating and DHW and also outlets for space heating. The pellet burner is fitted in the upper half and always gives priority to solar heating. The boiler is a complete solution built in one unit, which gives compact and cost-effective heating. The storage boiler gives great flexibility as the base heat source can easily be changed between electricity and pellets. Moreover the type of fuel used can be changed by installing a gas or oil burner if preferred*

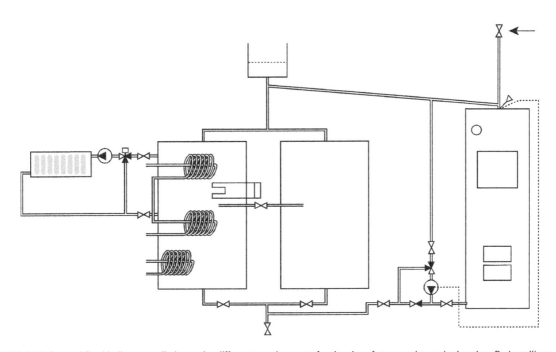

FIGURE 5.10 *A wood-fired boiler normally has quite different requirements for the size of storage than solar heating. By installing a master tank (with the technical equipment) and a slave tank, the conflict in volumes is resolved in a rational way. Thus the wood-fired boiler's requirement for a larger volume is met by using both the tanks, and only the master tank (on the left in the illustration) is used during the summer. An electric heater is the best way to guarantee the temperature and to avoid having to light the boiler during the summer*

5.4.6 KITCHEN BOILERS

A kitchen boiler burns most efficiently and with least environmental impact if it is connected to a relatively small storage tank. The kitchen boiler's heat capacity is normally under 10 kW, which makes it suitable for the DHW demand of most single-family houses (Figure 5.11).

A standard system with solar heating and a kitchen boiler should have 8–12 m² of solar collector and a storage tank of about 500 litres. The storage tank is equipped with an electric heater with the same power as the kitchen boiler (equal to the total power demand of the house). The electric heater works primarily as a temperature guarantee, but also has an important function as it obviates the arduous necessity of keeping the boiler alight. In principle the function of the system is that the kitchen boiler (or alternatively the electric heater) supplies basic heating during the autumn, winter and spring. In summer the system is completely

dependent on the sun and, if necessary, electricity. The time when the kitchen boiler is in use is minimized, and the solar collectors are utilized to the maximum. It is an interesting choice of system for smaller and newer single-family houses.

5.4.7 WATER-COOLED INDIVIDUAL HEATING UNITS

There are water-cooled (that is, water-jacketed) individual heating units (soapstone stoves, tiled stoves, closed stoves and pellet stoves) that have a similar function and system design to kitchen boilers. Water-jacketed heating units with a low heat capacity (1–5 kW) need a smaller storage volume. The solar heating will determine the size of the storage tank where applicable. Solar heating will show to advantage in combination with individual heating units as the DHW demand is guaranteed by the solar collectors during the summer period when the boiler is not lit.

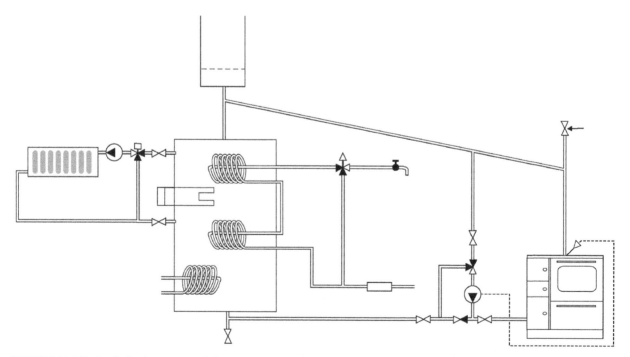

FIGURE 5.11 *A kitchen boiler does not usually have a very large heat capacity, and can therefore use the same storage volume as solar heating. The common storage tank should be equipped with an electric heater as a reserve. This type of system is of interest for well-insulated single-family houses where the heat load for space heating and DHW is not very large*

5.4.8 SOLAR HEATING AND HEAT PUMPS

The combination of solar heating and heat pumps is not yet particularly common. The reason is, above all, that both solar heating and heat pumps require relatively high capital investment. Furthermore, expert knowledge is required to be able to utilize the advantages of both types of energy and both systems in the same plant. Technical competence is required just to get both types of energy to operate together reliably and economically. However, it is interesting that solar heating and heat pumps both have the best output at low working temperatures.

Probably the most profitable combination is to let the solar heat reload the heat pump's heat store, for example a borehole or a heating loop in the soil. The system must be constructed correctly. To mention one example, a soil heating loop must never be charged at a temperature higher than +20°C. The input temperature for a heat pump (applicable to all compressors) must never exceed +20°C either. One way to increase the output of solar heat and increase the efficiency in the combined system is to keep the high-pressure side of the heat pump to approximately +40°C. For this to succeed it must be combined with floor heating and the DHW temperature must be topped up in a small, separate hot water tank. Always consult a specialist for correct sizing and system design, and check the rules and regulations in the relevant country.

Two possible types of combination can therefore be expected in the future, one with an exhaust air heat pump in new houses and one with ground (soil and rock) heat pumps in houses built during the 1970s and 1980s (see Figure 5.12). There is a need for flexible heating systems based on a storage tank for both of these target groups. A relatively low-powered exhaust air heat pump can make use of the heat content of the ventilation air, transfer it and transport it further to a storage tank. This small heat contribution is unlikely to disturb the solar heating performance.

A further option is to combine solar heating and a heat pump where the solar collector can raise the temperature in the ground coils in a ground (or rock) heat pump. At present there are few installations of this

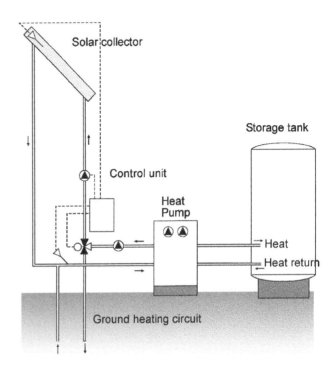

FIGURE 5.12 *In recent years several manufacturers have begun to combine ground heat (rock or soil) with solar collectors. Solar heating is utilized in the best possible way as the heat produced is utilized and given priority according to the best use for the temperature level. As soon as the solar heating loop has a sufficiently high temperature, DHW and/or space heating is produced. At lower temperatures the solar heat is fed into the heat pump. Excess heat in the summer, or low-temperature heat during the spring, winter and autumn charge the borehole. In this type of system solar heating can be used for a wide temperature range, resulting in large heat gains*

type, so the technique can be considered to be relatively untried. There are no technical or operational problems. Perhaps the future can offer system designs in which simple, low-temperature solar collectors are marketed as a complement to, or as a standard product with, ground heat pumps (or rock heat pumps).

5.4.9 CULVERT SYSTEMS

There are two types of culvert system. In one the house owner installs a heating system in an adjacent building separate from the house; then the culvert connects the boiler room to the rooms that are to be heated. The system should be classified according to the form of auxiliary heating. The other type is a culvert system from a local heating plant to separate properties, for example a small number of detached houses or terrace houses. Solar collectors can be connected to the system in different ways, either directly to the common plant or with individual solar heating systems for each separate property (Figure 5.13).

An individual system solution can be of interest even with these common plants – either as a safety measure, or to motivate the individual consumer to save energy. When this type of system is required, a small storage tank in every one of the connected households may be preferable. With a storage tank of 300–500 litres, part or all of the heating demand can be covered by water-jacketed individual heating units combined with solar heating, or solar collectors in combination with an electric heater. The main heat supply is via a culvert from the common plant. The heat from the culvert system is transferred to a storage tank in each house.

Local heating plants of this type make use of the advantages of large-scale production. Vulnerability is reduced with individual storage tanks at the same time, as the central boiler can be turned off for 4–6 months per year. A further advantage is that the individual house owners are partly responsible for the heating system.

5.4.10 DOMESTIC HOT WATER (DHW) SYSTEMS

Many makes of hot water tank have a heating coil installed in the lower part of the tank as standard. This can be used for connecting the solar collector (Figure 5.14). This system solution is mainly of interest for properties with electric resistance heating.

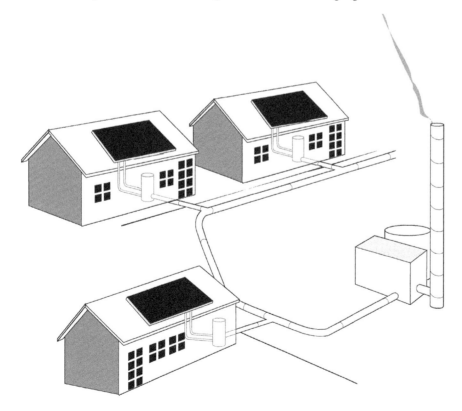

Figure 5.13 Individual heating systems connected to a small, local heating plant. The advantage of the system is that individual solutions further energy savings and personal wishes at the same time as the vulnerability of the system is reduced. A common heating plant with, for example, a wood chip boiler allows both lower production costs and minimal environmental impact during periods of high load

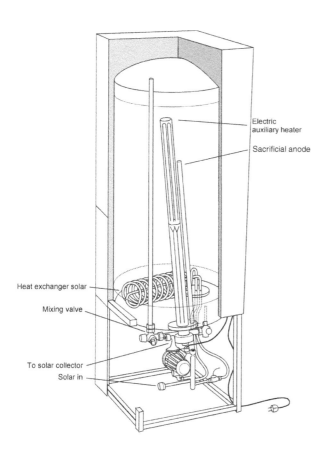

Electric
auxiliary heater

Sacrificial anode

Heat exchanger solar

Mixing valve

To solar collector

Solar in

Figure 5.14 There are a number of different suppliers of hot water tanks fitted with solar heating loops. It is important that the solar loop be placed low down in the hot water tank and that other auxiliary heating (usually an electric heater) is placed a little higher in the tank, so as not to impair the temperature stratification. The cold water inlet should be in the bottom, and the hot water should preferably be drawn off from the upper part of the tank. Choose the type of hot water tank with regard to the actual water quality. When the hot water tank has a sacrificial anode it must be renewed according to the supplier's instructions. Set the temperature of the electric heater accurately, among other things to hinder the growth of legionella bacteria

There are various makes of hot water tank, where the construction and choice of material are suitable for different water qualities. To avoid the risk of corrosion from the oxygen-rich, fresh water the hot water tank can be copper lined, enamelled or made of stainless steel. It is important to seek advice from local specialists on choosing the correct type of hot water tank to suit the local water quality.

Normally hot water tanks are 200–300 litres, which suits the solar heating system well. An area of 4–6 m² of solar collectors, connected to a tank volume of around 300 litres, covers 50% of a normal Swedish family's annual demand for domestic hot water. For maximum utilization of the solar heat it is important to place the solar heating coil as low as possible. The electric heater should be placed in the upper half of the hot water tank, which means that the solar loop always works at as low a temperature as possible. The manufacturer's instructions should be followed carefully, as well as any rules and regulations in the relevant country on the temperature setting for the hot water tank to avoid the growth of legionella bacteria.

In Sweden there has been a special type of tube-in-tube heat exchanger for existing hot water tanks (Figure 5.15). The solar heat can be transferred from the cold water inlet via a tube-in-tube heat exchanger – a relatively inexpensive installation. Solar heating can also be transferred to an existing hot water tank with a traditional flat-plate heat exchanger. However, the hot water tank must have inlet and outlet points for this.

There is a large target market – about 500,000 homes in Sweden – for DHW systems in single-family houses with electric resistance heating when the hot water tank is replaced. The standard system, which has been made more cost-efficient in connection with a technical competition, comprises a hot water tank with solar coil (external dimensions 60 × 60 cm, 250–300 litres), all necessary equipment, and a solar collector of 4–8m². It is now available on the market at a favourable price.

5.5 Operation and maintenance

Regular supervision and attention are necessary for the solar heating system to work well. It is important that the service and maintenance instructions are available.

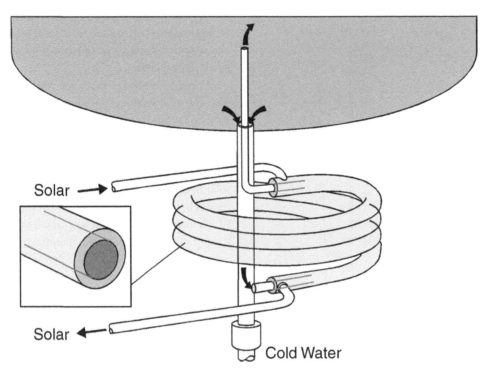

Solar →

Solar ←

Cold Water

FIGURE 5.15 *In the middle of the 1980s a special type of tube-in-tube heat exchanger was developed for connection to existing hot water tanks. To be able to connect the heat exchanger the cold water inlet must be at the bottom of the hot water tank. Depending on the properties of the heat transfer fluid used, the installation of this type of heat exchanger must be checked for approval by the relevant authority in the country*

Make sure that the supplier provides manuals, data sheets and similar information on delivery

Some general hints on operation and maintenance of a solar heating system for single-family houses are given in Boxes 5.1 and 5.2.

5.6 Summary

Single-family house owners have good opportunities for using solar heating. A common requirement for all uses is a heat store, partly to even out temperatures and partly to store solar heat. It can be a hot water tank, a storage tank, or an outdoor pool. In the future, other forms of storage may be possible, for example in salts or in ground storage, where the solar heat is stored in the ground to be used in combination with a heat pump.

Solar heating is a good complement for the single-family house owner, and gives great opportunities for reducing heating costs. For a normal family of four persons, 4–6 m^2 of solar collectors are enough to cover about half of the DHW demand. When solar heating is connected to the heating system and is expected to contribute to space heating the solar collector area is increased to 8–12 m^2, which in turn determines the storage volume. Normally 50–100 litres storage volume is recommended for every m^2 of solar collector.

The solar heating system should be designed for the summer load – that is, the demand during the most intensive solar period. Thus the area of the solar collector is often determined primarily by the DHW consumption; the rest of the heating system and the auxiliary heating ensure that the demand is met during

Box 5.1 Operation and components in a solar heating system for single-family houses

■ The heat collected by the solar collectors is shown by the rise in temperature in the heat store. The amount of heat stored can be calculated by noting the rise in temperature at different levels in the tank (lower, middle and upper parts of the tank) after a day's solar radiation. Based on the rise in temperature in the tank, ΔT, the heat increase (in kWh) can be calculated using the equation $Q = 1.16 \times V \times \Delta T$ (where V is the volume of the tank in m³). Better, more exact measured values are obtained if heat consumption is minimized on the day of measurement.

■ The rise in temperature between the supply and return from the solar collector should correspond to the values given by the supplier. This can be governed by the pump speed and the flow valve setting.

■ By feeling the cover glass on the solar collectors it is possible to decide whether the output is uniform. A good indication of the correct flow is if the cover glass feels cold. When several solar collectors are coupled in parallel the temperature on the cover glass should be the same, and for coupling in series the last solar collector should be hotter than the first collector in the loop.

■ When heat input to the storage tank is suspiciously low the capacity of the heat exchanger should be checked.

■ If there is air in the solar collectors it can be clearly heard as an uneven sound in the pump.

■ Over and above faulty connections and material faults, operational disturbance in most cases is caused by too low a flow rate. Disruption in flow is normally caused by air in the system, by a valve, or by another component being clogged.

■ Regular checking of the pump's start and stop function is recommended.

■ Check the level of fluid and the pressure in the solar loop at regular intervals, always in connection with or after air bleeding has been carried out.

■ Check that the pipe parts and component parts are insulated correctly: this is especially important for pipe parts outdoors.

■ If the function of the control unit is in question it is advisable to check the position of the sensors and that the temperature sensors are carefully insulated.

Box 5.2 Annual maintenance of solar heating system for single-family houses

■ Inspect the fixing of the solar collectors (and the flashings around the solar collectors where applicable) regularly.

■ Check the pressure and level of the liquid in the solar heating system at least twice a year. If the level of liquid or the pressure goes down, suspect leakage.

■ Check the possible need for air bleeding of the solar loop at least once a year.

■ Check the start and stop function of the solar heating control unit and also the position and insulation of the sensors.

■ Check the operation of the pump.

■ Inspect the connections and joints in the system; some heat transfer fluids have a great propensity for creeping, which cannot always be registered by a pressure gauge or levelling vessel.

■ Insulation material is usually a popular nesting material for birds and rodents.

■ Clean the filter and/or non-return valve annually. Pieces of solder from the copper pipes often stick in these components and affect the flow in the solar loop.

■ Check that the rise in temperature in the solar loop and the storage tank comply with the supplier's recommendations.

■ In systems that use glycol mixed with water as the heat transfer fluid, check the freezing point annually. Furthermore, make a chemical analysis of the glycol-mix fluid at regular intervals to check the corrosiveness of the mixture. Change the liquid or add an inhibitor when necessary.

■ Regardless of the heat transfer fluid in the solar loop, check the freezing point of the heat transfer fluid annually and the chemical composition regularly (e.g. every third or fifth year or according to the supplier's instructions).

the high load period and therefore affect the design of the heating unit to a greater extent. Normally the basis of the system design is a storage tank to which both the solar and other heat producers deliver heat.

As solar heating often demands a storage tank, the flexibility of the heating system increases. The storage tank provides a number of different options for combinations of systems where solar collectors can be combined in the best way (both technically and economically) with all sorts of energy sources. Solar heating has a competitive advantage in combined systems with solid fuel (biofuels) and fluid fuels (oil and gas). During periods of low load, when solar heating can, in principle, cover the whole DHW and space heating supply, the auxiliary boilers have the worst efficiency.

There is a large selection of solar collectors on the market. Flat-plate solar collectors still lead the market. They are often marketed as complete prefabricated units, made to measure and for a number of different roofing materials. In some countries there are semi-manufactured units that are assembled by the customers themselves. Evacuated solar collectors are another type that are increasing their share of the single-family house market.

■ 6. Solar heating for outdoor pools

6.1 Energy and heat data

The need for heating in an outdoor pool is determined by the desired water temperature. The heat needed is the difference between the heat input from solar radiation on the water surface and the heat losses (evaporation and convection). At Swedish latitudes without energy-saving measures the heat losses can reach 1200 kWh per m² of pool area during the operating season. Therefore temperature control and the pool covering are important to reduce the need for extra heat.

An outdoor pool without energy-saving measures in Sweden has a net energy demand (after passive solar contribution by direct absorption by the water) corresponding to 700 kWh per m² of pool area per year (season) with a temperature demand of 25°C. For a normal 50 m outdoor pool this means a net energy demand of 420,000 kWh during an operating season. The corresponding net energy demand for a private outdoor pool (standard pool 8 × 4 m) is slightly over 20,000 kWh per year.

The heat demand for an outdoor pool varies according to the temperature requirement and climate factors (wind, outdoor temperature and solar radiation), and whether there is a pool cover. If the pool is consistently covered, the energy consumption can be reduced by up to 40%. If the pool is placed in a sheltered area, a further 30% of the heat demand can be saved. An outdoor pool in the north of Sweden, around the Arctic circle, does not necessarily need more additional heat than a pool in the south, around latitude 56° N, in spite of less favourable conditions with lower temperatures and less solar radiation, because of such factors as different average wind speeds during the year, patterns of use, and system design.

Solar insolation on the ground, between latitudes roughly 56° and 65° N is from slightly under 500 kWh/m² to slightly over 600 kWh/m² during the period from May to August, depending on location. Direct absorption by the water and the adjacent material amounts to about 300–500 kWh/m², except for pools less than 1 m deep, for which absorption by the water will be lower. In both cases it is possible to make use of passive solar energy by using coloured pool material such as ceramic tiles.

A solar collector area (simple unglazed solar collector) that is 50–100% of the pool area can produce heat corresponding to 200–400 kWh/m² per season in Sweden: that is, more than half the energy consumption in an ordinary pool where the temperature requirement is at least 25°C (see Figures 6.1 and 6.2). For a pool with an area of 1000 m² this means a heat production of 100–200 MWh, which gives an energy cost of 65,000–130,000 SKr (€7,222–14,444) per season, with today's (2002) energy prices. For unglazed solar collectors in a system of this size the payback time will be 5–10 years depending on interest costs, period of amortization, alternative energy prices, and changes in these. This is on condition that there is a pool cover and that other energy-saving measures are carried out before the solar heating plant is sized.

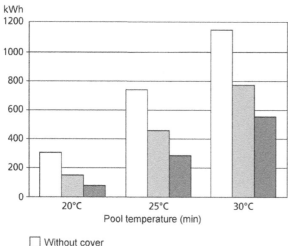

FIGURE 6.1 *Energy consumption during the period May–August in an outdoor pool in Stockholm with 1000 m² pool area. The consumption varies greatly with the pool temperature. If the pool cover is used, the amount of energy used is reduced considerably Source: Svenska Solenergiföreningen (1992)*

Running and maintenance costs for a correctly constructed solar heating plant for outdoor pools are very low. Energy for running the pump is only a few percent of the annual energy output.

6.2 Direct systems: unglazed solar collectors

There are great differences between direct and indirect solar heating systems for outdoor pools. In an indirect system (see section 2.7.1) both the pool water and the DHW are heated by the solar system (see Figure 6.8). However, in a direct solar heating system the solar heat is transferred to the pool water directly in the solar collector. No heat exchanger is necessary as the chlorinated pool water is used as the heat transfer fluid (Figure 6.5). The system design is simple. The pool water is led up to the solar collector directly after the filter in the pool circuit. In smaller systems (for private houses) the pump in the pool circuit is normally used

FIGURE 6.2 *Unglazed solar collectors in a direct system for heating an outdoor pool*
Photo: Lars Andrén

for circulation through the solar collectors as well. In larger systems an extra pump is often needed for the solar heating system (see Figure 6.6(A)).

In an unglazed solar collector the absorber material is plastic (such as UV-stabilized polyolefin) or some sort of rubber material (such as extruded EPDM rubber), which can stand chlorinated water. An advantage of rubber absorbers is that they can stand temperatures below zero (the solar collector can expand). However, the suitability of the material in the pipes should be checked. The unglazed solar collector is affected completely differently from a glazed one by the ambient temperature, wind load and working temperature.

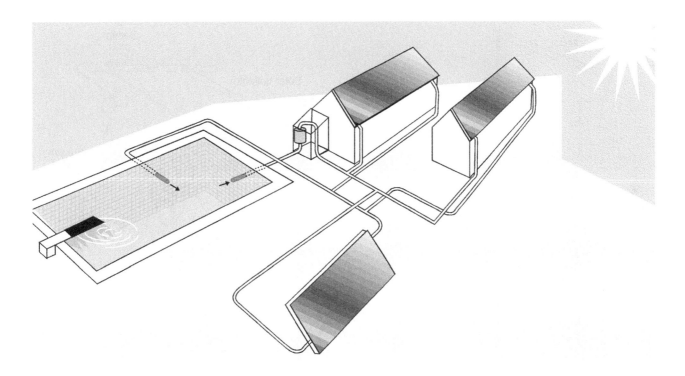

FIGURE 6.3 *Solar collectors for outdoor pools are normally sized according to the area of the pool. In Sweden the solar collectors normally should be 50–100% of the pool area, depending on the temperature requirement and wind-cooling effect*

Unglazed collectors are suitable only for systems with a low temperature demand. The efficiency of an unglazed collector falls considerably with an increase in working temperature in the solar loop (see Figure 6.7). It can often be an advantage to place an unglazed solar collector almost horizontal (however, follow the supplier's instructions regarding the smallest tilt), not least considering the angle of incidence of the sun during the actual period of operation.

Box 6.1 lists some hints for unglazed solar collectors in direct systems for outdoor pools.

6.2.1 OPERATION

Small and medium-sized systems are based on existing equipment. Connection to and from the solar collectors is by two T-pieces (with a stop valve between them), fitted after the cleaning equipment and before possible auxiliary heating. If the pump in the pool circuit has sufficient capacity it is also used for circulation through the solar collectors. Be careful not to disturb the flow through the filter system. It is important that all the components in the system can stand the prevailing outdoor conditions (such as UV radiation, humidity and temperature) as well as the chlorinated water. Polymers (polypropylene, polyethylene and EPDM rubber) of the right quality will normally fulfil the requirements, but metals are less suitable because of the risk of corrosion. Always check with the supplier or another expert.

The working temperatures in the solar loop are relatively low and seldom need to rise over 30°C. The temperature differences between the feed and return are small, and can be between 0.5 and 3°C depending on the type of solar collector and recommended flow. This places heavy demands on the rate of flow (which must be correct and constant) and the accuracy of the control equipment.

Direct systems with unglazed solar collectors need a guaranteed drainage function. There are rubber solar

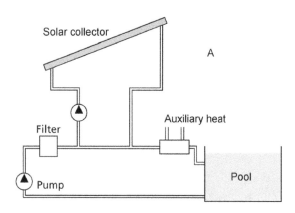

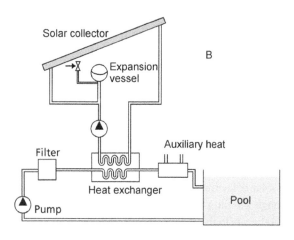

FIGURE 6.4 *(A) Direct system with unglazed solar collectors; (B) indirect system with glazed solar collectors. When an extra pump is fitted for the solar loop this can be directly on the supply pipes to the solar collectors, as in (A). Alternatively the pump can be fitted in a closed solar loop, which must then have a separate expansion vessel, as in (B). The latter system can be used for operation throughout the year, provided the heat transfer fluid in the solar loop is freeze-resistant*
Source: Svenska Solenergiföreningen (1992)

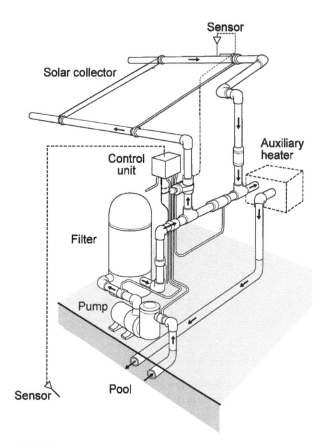

FIGURE 6.5 *In direct solar heating systems for outdoor pools, the chlorinated water is used as the heat transfer fluid and flows directly through the solar collectors. In smaller systems for single-family houses the existing pump can normally be used. Note that possible auxiliary heat is connected after the supply pipe from the solar collectors*

collectors that can stand being frozen, but all types of plastic solar collector and plastic piping will burst at the least touch of frost if they are full of water. The drainage function is very important, and should be designed so that it takes place automatically as soon as circulation stops in the solar loop.

Normally the solar collectors are placed higher than the pool surface, which can cause negative pressure near the outlet from the solar collectors. This can make draining more difficult and, by taking in air, interfere with the circulation in the pool. Box 6.2 lists some hints for safer drainage.

A check valve on the return pipe can give slight over-pressure in the solar loop. This makes operation more reliable and allows a simpler construction of the air bleeder and the air inlet on the highest point of the system. Then venting and air intake (on draining) can

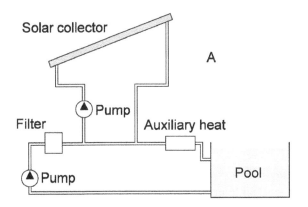
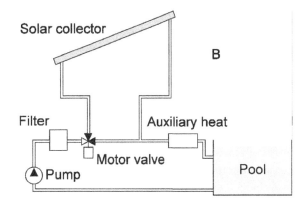

FIGURE 6.6 *(A) Direct system with a separate pump connected via the filter circuit; (B) direct system with common pump for the solar collector and filter circuit. A three-way-valve controls when the pool water should be passed through the solar collectors*
Source: Perers (1992)

Box 6.1 Hints for unglazed solar collectors in direct systems for outdoor pools

▨ Use insulation of polymer material that can stand temperatures up to 90°C.
▨ When fitting pipes, work from the lowest to the highest point. In this way drainage and air bleeding are made easier.
▨ Install facilities for drainage and air bleeding at low and high points.
▨ Install shut-off valves to make service easier (for example when changing the pump, or cleaning the filter).
▨ All material that comes into contact with the pool circuit must be resistant to chlorinated water.
▨ Connect the components in the right order. Connect the solar collectors after the filter but always before an auxiliary heat source.
▨ Pipes under the ground have to stand traffic loads.
▨ Test the pressure before insulating and starting up.

Box 6.2 Some hints for safer drainage

▨ Air bleeders on the highest point of the solar loop.
▨ Sloping main pipes from the highest to the lowest point in the system (pool or collection vessel). Also applies to the pipes and connections in the solar collectors.
▨ A non-return valve prevents the drainage flow from going backwards through the filter in the pool loop.
▨ Complete drainage can be affected by pressure drop or static pressure drops in the filter circuit.
▨ It should be possible to check the drainage function simply and regularly.
▨ Shut-off valves should keep the draining parts of the system free from liquid during long shut-down periods, without disabling the pressure valves.
▨ Follow the supplier's directions carefully.

---

be carried out with an open glass gauge on the high points of the system. This type of system is also a protection against excessive over-pressure in plastic and rubber absorbers. Water flow rates of slightly over 0.5 m/s make venting easier. Always follow the advice and instructions from the relevant supplier or other experts. Consideration must be paid to pressure drop and the static pressure drop in plastic solar collectors

(polymer, polyolefin and so on). Avoid excess pressure larger than that given by the manufacturer.

6.2.2 SOLAR COLLECTOR LOCATION
The location of an unglazed solar collector for summer use does not have to fulfil the same requirements as a glazed solar collector for all-year operation. The determining factors are orientation and the amount of overshadowing. The angle of the solar collector to the horizontal plane is not so important for an unglazed

solar collector that is used only during the summer, primarily because the collector does not have a glass cover that absorbs and reflects the solar radiation.

While an optimal tilt for a glazed solar collector is 10°– 15° below the latitude in northern Europe, a lower tilt can be recommended for summer use (such as heating outdoor pools) and a steeper slope for year round use. Variation in annual output at different angles is surprisingly small.

Deviation from a southerly direction also has a relatively limited effect as long as the solar collector is not placed facing directly west or east. In drainback systems the tilt of the solar collector should not be less than 15° from the horizontal plane. Carefully follow the directions from the supplier so that the drainage works correctly.

All forms of overshadowing must be avoided. Narrow objects such as ventilation pipes and flagpoles have little effect (but must be avoided with regard to shadowing of solar cells). A rule of thumb to avoid overshadowing during the summer is that the minimum distance from the solar collectors to an object on the south should be 1.4 times the height of the object. When the object is on the south-east or south-west the factor is 1.8 and directly to the east or west it is 4. These figures are for latitude 60° N but the distances can easily be calculated for other latitudes.

For correct signals to the control equipment, and hence for correct control of the system, the temperature sensor for starting and stopping the plant should be placed free from overshadowing. In the same way, the solar collectors' supply and return piping should, as far as possible, be placed so that it does not directly affect the temperature sensors in the pool water – that is, the temperature sensors for the solar loop and a possible temperature sensor for the auxiliary heating. In some cases the sensor is on the main pipe for the pool water in the control room: if so, it should be well insulated so that it is not affected by the room temperature.

A fence that gives shelter or vegetation as protection against the wind can be of great advantage for the efficiency and heat output of unglazed solar collectors. If the solar collector is placed directly on the ground or a roof surface the absorber is automatically protected.

Locating the solar collectors on roof areas next to the control room simplifies pipe runs. However, it requires good waterproofing around the fixings and possible pipe runs through the roof. This is particularly important on flat roof surfaces. It is also important to analyse the compatibility of different materials, so that the solar collectors do not corrode roofing material, for example. Wind and snow loads must also be considered.

6.2.3 SIZING

A constant temperature cannot be guaranteed in a pool that is warmed only by solar heat. As the main heat loss is through evaporation and convection from the surface of the pool, it is the area of the pool (and not the volume of water) that is the basis for sizing. A rule of thumb for unglazed solar collectors is that the solar collector area should be equal to 50–100% of the pool area. The solar collector area should not exceed 1.5 times the pool area, to avoid the production of excess temperatures; note that the growth of bacteria increases greatly with pool temperatures over 28–29°C. Follow the supplier's instructions with regard to local conditions.

Normal calculation formulae can be used to size the main pipes to and from the solar collectors. An even distribution of water through the solar collectors (groups) is desirable, and the flow rate in the system should reach slightly over 0.5 m/s. This is to obtain a low pressure drop at the same time as the air in the system follows the flow. By keeping a low pressure drop the power of the pump can be limited to 5–10 W per m^2 of solar collector.

The flow through unglazed solar collectors should be as large as possible, within the manufacturer's recommendations. An unglazed solar collector has a heat output that is extremely temperature dependent; a flow of between 2 and 4 litres/min per m^2 of solar collector can be recommended (for glazed solar collectors 1–2 litres/min per m^2). Too small a flow gives less heat output as the solar collector's running temperature will be unnecessarily high, not least because the pool does not have any temperature stratification to speak of.

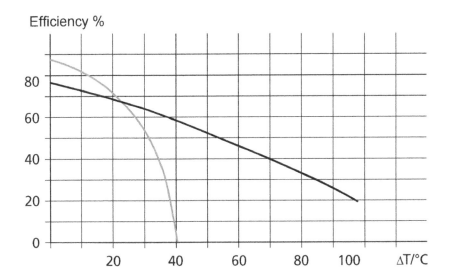

Efficiency %

FIGURE 6.7 *An unglazed, uninsulated solar collector can easily be more efficient than a sophisticated, well-insulated type. When the working temperature in the solar loop is close to the outdoor temperature, and the period of operation covers the warm summer months, an unglazed solar collector is more efficient as there are no losses due to reflection or absorption in a glass cover and also because of the favourable working temperatures*

Box 6.3 sets out a checklist for purchasing, and Box 6.4 gives an example of a sizing calculation for central Sweden.

There is no reason to invest in highly sophisticated control equipment for the solar loop. However, the control equipment must be sufficiently accurate as the temperature differences in the system are often low,

normally 3–5°C. Time control of auxiliary heating can be of interest, for example to make use of differentiated electricity tariffs.

Equally, large, parallel-coupled solar collector areas give an even flow distribution in the solar loop. A certain amount of self-regulation of the flow is achieved by the inner friction of the water being greatly reduced with rising temperatures, which results in the pressure drop being reduced in the solar collector. Good deaeration is an important condition to achieve good flow distribution.

The better the weather the greater the number of bathers and vice versa. For this reason a large part of the auxiliary heat can be saved if the pool temperature is allowed to 'follow the weather'. When the weather is poor the pool temperature is allowed to sink. With good weather the solar collectors and the passive heat gain (solar radiation direct on to the pool surface) automatically raise the pool temperature.

6.3 Indirect systems: glazed solar collectors

Glazed solar collectors in indirect systems are used only where the solar heating system has a load throughout

---

Box 6.3 Pool heating: checklist for purchasing

- Survey the prerequisites.
- Design and size, taking into account requirements and special conditions.
- Prepare drawings and specifications, tender documents (for possible site work, pipelaying, electrical work, building works).
- Apply for building or planning permission and other necessary permits where applicable.
- Obtain prices based on tender documents.
- Determine form of contract.
- Go out to tender (include guarantee conditions).
- Final inspection and control of operation by impartial inspector.
- Compile detailed documentation. This applies to the whole process from the initial survey to the operation and maintenance documents and final inspection report.

Box 6.4 Example of sizing calculation for central Sweden

Outdoor pool facility:	Main pool 50 m × 12 m, children's pool 16 m × 6 m (total volume 1500 m³)
Present consumption:	60 m³ oil per year (corresponds to 500,000 kWh of electricity)
Period of operation:	The pool facility is open from 1 May to 30 August
Suitable solar collector area:	Approximately 520 m² unglazed
Solar radiation:	May 125.55 kWh/m²
	June 127.42 kWh/m²
	July 122.55 kWh/m²
	August 92.75 kWh/m²
	Total: 468.27 kWh/m² during the bathing season
Total energy:	520 m² × 468.27 kWh/m² = 243,500 kWh
Average efficiency:	70%

Calculated energy saving: approximately 170,000 kWh/season (70% of 243,500 kWh), corresponding to 34% of the total heat consumption.

---

the year. This can be in combined pool facilities where there are both indoor and outdoor pools, or in facilities where an outdoor pool is next to some sort of winter use, such as a sports hall. A glazed solar collector with an antifreeze heat transfer fluid will give additional heat throughout the year. If the solar heating is sized for an outdoor pool, there is also a large amount of additional heat during the spring and autumn (when the pool is not in use), and it is important to take this into consideration.

Indirect solar heating systems are based on conventional heating, ventilation and sanitation technology, and have system components similar to those in other solar heating systems with glazed solar collectors. It is important to study the sizing prerequisites carefully.

6.3.1 SOLAR COLLECTOR LOCATION

Glazed solar collectors for all-year operation have requirements for position, orientation and tilt from the horizontal plane that are different from those of unglazed collectors for summer-only use. In Sweden it is recommended that a glazed solar collector be orientated from south-west to south-east with a tilt of 10°–15° below the latitude. Deviation from this gives a loss in heat production of less than 10% between the best and worst cases: see Table 2.1.

It is always an advantage to place the solar collectors as near to the control room as possible to minimize costs and heat losses in the pipe runs. If the solar collectors are used for an outdoor pool a good position is on a roof. When they are placed on the ground there is always the risk that they will be damaged, or visitors may be injured. In most cases there is space available for the solar collectors; do not let a lack of imagination limit the options.

Box 6.5 lists some points to bear in mind when positioning the solar collectors.

6.3.2 DESIGN AND SIZING

Sizing should be based on the solar collectors' producing enough heat for the existing heating to be turned off during the summer period. This is particularly important when the auxiliary heating is from an oil-fired boiler. The ability to turn the oil-fired boiler off during the summer period leads to lower costs for oil and reduced environmental impact. Figure 6.8 shows an indirect design in which part of the solar heat is used to produce DHW before the remainder is used to heat the pool water.

The solar heat should not be designed to keep a constant pool temperature. When a minimum temperature is required for the pool water it is necessary to have an auxiliary heating source.

The size is always based on the summer load – that is, the required pool temperature. The area of the solar

Box 6.5 Points to remember when positioning solar collectors

- Avoid all forms of shadowing; watch out for vegetation, and objects on the roof that can cast shadows.
- Be careful with roof penetration, in particular in flat felt roofs.
- Check whether building, planning permission or other permits are required. Check for other possible local or central government regulations. Follow the regulations regarding connection to water and sewage services.
- Services in the ground must stand traffic loads when applicable, and there must be drainage points on the low points of pipe runs.
- Pay attention to relevant standards for snow and wind loads when placing on a roof.
- Make sure that there is space for service around the solar collectors.
- Make sure that the heat transfer fluid is collected after control of leakages or after draining the solar loop.
- The solar collectors and the control room should be as near to one another as possible.
- Check that the solar collectors do not shade each other.

collectors is based on the area of the pool, temperature requirement, heat losses, the efficiency of the solar collector and local conditions. Winter consumption is of minor importance. However, it is important that the volume of the store takes into account the heat that the solar collectors produce during the periods in the spring and autumn when the outdoor pool heating is not in operation and the heat is to be used elsewhere. This should be considered as early as at the design and sizing stage.

The pipes are sized according to normal calculation formulae. An even flow distribution with a reasonable pressure drop is created by the solar collector area being divided into parallel-coupled groups of equal size. Control valves are not normally necessary to adjust the flow distribution, as the pressure drop through the main pipes and the solar collectors is seldom too large or too variable.

Be careful when choosing the material for pipes (including all pipe parts and valves) that come into contact with the chlorinated water. Copper is

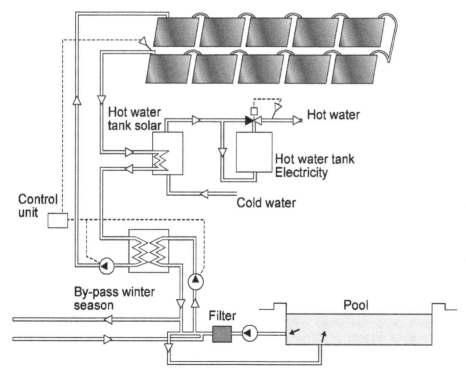

FIGURE 6.8 *An indirect solar heating system can be extended from heating pool water only to producing DHW as well. Glazed, well-insulated solar collectors have output temperatures that can be successfully used for DHW production. Shown here is a system design in which part of the solar heat is used for DHW before the remainder is used to heat the pool water. Solar collectors can produce hot water throughout the year. It is important that there is a use for the solar heat during the spring and autumn when the outdoor pool is not in use*

particularly vulnerable when the pH value is too low, and acid-resistant stainless steel pipes may have limitations with regard to high working temperatures. Corrosion is commonly caused by the use of a combination of different metals in the loop, as a result of galvanic corrosion caused when different metals are in electric contact with one another. This can happen when there is a leakage of oxygen into the heat transfer fluid.

6.4 Key ratios for energy and economy

To maintain an average temperature of 25°C, about 700 kWh per m² of pool area is needed in Sweden during the summer season from June to August. This gives a heat demand of approximately 420,000 kWh for a standard pool of 50 × 12 m (that is, a pool area of 600 m²) during the summer season. With an energy price of 0.65 SKr (€0.07) per kWh the seasonal cost for heating the pool will be about 273,000 SKr (€30,333).

Test results from SP show that the annual output from unglazed solar collectors in Sweden varies between 200 and 400 kWh/m² per year at 25°C working temperature in an outdoor pool. If 450 m² of solar collectors are installed (equalling 75% of the pool area), the saving in cost will be between 58,000 and 115,000 SKr (€6,444 and €12,778): thus the heating costs can be halved. The payback time is about 5 years, depending on the type of financing and changes in energy prices.

In Falkenberg, south-west Sweden, there is a solar heating plant with 125 m² of glazed solar collectors for pool heating that has produced over 400 kWh/m² annually for about 10 years (125 m² × 400 kWh × 10 years = 500 MWh) (see Figures 6.9 and 6.10). To obtain a good yearly output from the solar collectors it is important that the winter production can be made use of in an energy-efficient way, which is the case in this plant.

For making a rough estimate and an analysis of profitability, see Box 8.3.

FIGURE 6.9 *Vessigebro swimming pool near Falkenberg, Sweden, has had high-temperature solar collectors for heating since the summer of 1985. The solar collector plant with 125 m² of solar collectors has catered for the greater part of the pool's heating and DHW demand during the bathing season (May–August). The annual saving corresponds to 15–17 m³ of oil, and the solar collectors have produced 400 kWh per m² of solar collector annually during the first 10 years of operation*
Photo: Lars Andrén

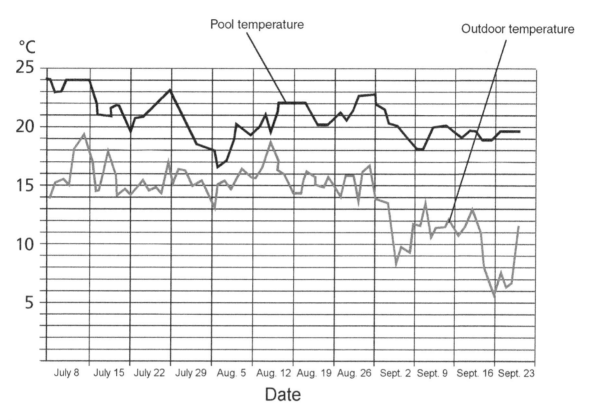

FIGURE 6.10 *Temperature graphs, Vessigebro swimming pool, Falkenberg. During the first summer in operation (the measurements started 8 July and continued until 23 September) the 125 m² solar collectors managed to maintain the pool temperature (total pool area 250 m² with a total pool volume of 350 m³) at approximately 5°C over the outdoor temperature. The bathing season can be lengthened during both the autumn and the spring with the help of the solar heating system, without auxiliary heating*

6.5 Operation and maintenance

The functioning and long-term operation of a solar heating system often depend on the service staff who have to maintain the plant – normally the regular maintenance staff at the pool facility. Good basic knowledge of the system technology is necessary for smooth running and maintenance. Therefore the supplier and the contractor must go through the plant carefully and describe and explain the function of the various parts that need annual inspection. Box 6.6 lists the basic contents of the operation and maintenance instructions.

A general rule when starting up a solar collector plant with an indirect system is that it should preferably be put into operation when it is not exposed to full sun.

On full solar insulation the solar collectors become very hot, which can cause flow problems and, in the worst case, boiling (steam build-up). As far as possible the solar loop should be filled in poor weather or when the direct solar radiation is cut off by the horizon. The solar collectors can, of course, be covered by tarpaulins if the area is not too large. The solar loop should be pressure-tested before start-up to look for possible leaks. The pressure test should be carried out with minimum solar insolation so that the pressure values are not affected by expansion due to temperature change. Note that unglazed solar collectors must *not* be pressurized. Tests for leaks must be carried out in some other way, for example by using the circulation pump when filling them with water.

Box 6.6 Pool heating: operation and maintenance instructions

■ Description of operation.

■ General terms of delivery for respective countries.

■ Data sheets and guarantee conditions for the components in the system.

■ Instructions on start-up, air bleeding, and normal running.

■ Routines for shut-down and freeze protection.

■ Routines for fault detection and recommendations for remedies for normal faults.

■ Instructions for manual operation.

■ Description of operation, and technical diagram of the control equipment.

■ Checklist for annual control and service.

■ Complete set of manufacturers' brochures.

■ List of addresses of consultants, contractors and suppliers.

■ Telephone number of personnel on duty, if there is one.

■ Safety regulations (for example to avoid boiling in the solar collector, on the properties of the heat transfer fluid etc.).

■ Date of installation.

■ Guarantee period and conditions.

Check the normal flow in the filter loop and solar collector loop, deaeration, leaks, flow direction in the solar collectors, possible drainage function in the solar loop and the operation of the control system on, or in connection with, start-up. All solar heating systems should be handed over fully adjusted by the supplier, if the handover takes place during the operating season.

In indirect systems the antifreeze mixture and the possibility of air in the system must be checked. The quality and the freezing point of the heat transfer fluid in the solar loop must be checked regularly. It is also important to check the temperature and flows on both sides of the heat exchanger.

A solar heating plant for an outdoor pool with a direct solar heating system must be turned off during the winter to avoid frost damage. The responsibility for this cannot be left open. Drainback systems (which drain automatically when the circulation pump stops) are advantageous here.

The pool filter system must be described carefully for the owner so that no misunderstanding can arise regarding the system's operation.

A solar heating plant that is correctly installed and is adjusted according to the supplier's directions is likely to operate without complaint. Wear and tear is minimal, and the system – which has few moving parts – should operate reliably for many years.

6.6 Summary

Heating costs for outdoor pools can be more than halved with solar collectors. The private home owner has a good opportunity to eliminate all the heating costs and at the same time extend the bathing season by some weeks, during both the spring and the autumn. In Sweden most plants make a profit after as little as five years.

A strong argument for solar heating as the heat source for slightly larger outdoor pools in Sweden is that these are normally heated by an oil-fired boiler with poor summer efficiency. Here the solar collectors will efficiently replace oil-fired heating, which will otherwise cause large emissions of carbon dioxide, sulphur dioxide and nitric oxides.

For pool facilities where only the pool water is to be heated, a simple unglazed solar collector is sufficient. The chlorinated water can flow directly through the solar collectors. If DHW production is also required, glazed solar collectors are used in an indirect system, in which the solar heat is transferred to heat both the chlorinated water and the DHW. In indirect systems a freeze-resistant heat transfer fluid is used, which means that the solar heating system can produce heat throughout the whole year.

It is important for the customer to specify requirements regarding temperature levels, operation, guarantee and service. The normal recommended water temperature is 25°C. The temperature requirements together with the local conditions and the total pool area determine the size of the solar collector area. A rule of thumb for colder climates is that the solar collector area usually equals 50–100% of the pool area.

Thus an outdoor pool with an area of 500 m^2 needs a solar collector area of about 250–500m^2.

It is important to have the right basis for procurement. Always make a survey of the existing conditions. Preliminary studies or design should always be the basis of the procurement. Check whether a planning or building permit is required for this type of installation. The relevant municipal authorities, utilities and energy companies should be contacted at an early stage.

Possible damage to roof coverings, gutters, walls and pipe insulation must be noted so that new damage can be made good by the solar energy contractor under the terms of the contract. Follow the general terms for delivery and any regulations in force in the relevant country. It is important to insure the solar heating plant/system, both during construction and after it has been put into operation.

Different tenders can easily be compared using a specification of requirements and responsibilities for installation of electricity, water, start-up, operational control, service agreement, and guarantee conditions.

■ 7. **Large-scale solar heating technology**

Large-scale solar heating technology is primarily intended for connection to district heating plants, DHW production for hospitals, local district heating plants for housing areas, or similar systems. There are two basic systems: one in which the solar heating system is designed to cover about 10% of the connected heating load (a *short-term heat store*) and one in which solar heating can cover up to 70% of the connected heat load (*seasonal heat storage*). In Sweden there are almost 1000 heating plants with yearly heating loads of 1–100 GWh for which solar heating can be a useful supplement. Solar heating plants with short-term storage can be an economic alternative to oil, particularly for local heating plants of between 7 and 32 GWh/year. According to a report published by NUTEK (the Swedish Business Development Agency) there are over 250 heating plants in Sweden with a total oil consumption of 2.7 TWh/year. Smaller local heating plants for housing areas where solar heating can be combined with biofuel via a short-term store are also of interest.

In recent years special large-module solar collectors have been developed for this type of solar heating. In certain types the large modules are 12–16 m² per unit. Other manufacturers build the solar collectors on site in complete lengths where the size of the unit can be adapted. The solar collectors are placed mainly on the ground, but are also to be found on roofs (Figure 7.1). Large-module solar collectors generally achieve a higher efficiency than solar collectors for single-family houses.

7.1 **Design and calculation**

The design of a large-scale solar heating plant is not appreciably different from that of other solar heating design: solar heat can, in principle, cover 0–100% of a normal heating load. Plants with a low coverage are normally uninteresting for many reasons, and plants with too high a coverage (over 80–85%) are unrealistically expensive. The design of a solar heating plant is based on the amount of heat that is required and not, as is normal for other heat sources, on a certain guaranteed heat output to cover the demand on the coldest day.

Solar heating can also be used to increase the total efficiency of the system, and this is often overlooked. For this reason the solar collectors' main task can be to reach a certain coverage or to replace all fuel combustion during a certain period, for example the low-load period during the summer.

7.2 **Short-term heat stores**

When a short-term heat store is chosen an annual coverage of 5–15% is generally reached, depending on climate and current heat load. In this type of solar heating system about 0.3 m² of solar collector is needed per MWh of annual heating requirement and approximately 0.1 m³ storage volume of water per m² of solar collector (Figure 7.4). If the solar collectors are oversized, the annual heat output can increase by

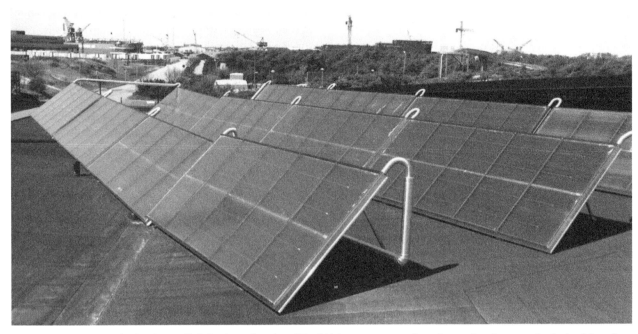

FIGURE 7.1 *Klitterbadet (swimming pool), Falkenberg, Sweden: 150 m² high-temperature solar collector placed on a gently sloping felt roof. Extra care is required when fixing solar collectors to a felt roof*
Photo: Lars Andrén

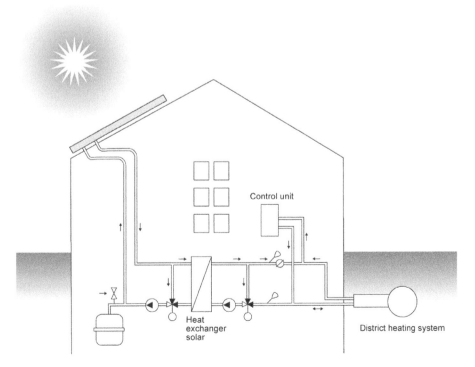

FIGURE 7.2 *This system transfers the heat from solar heating directly to the district heating culvert and utilizes the volume in the circuit as a heat store, which reduces the construction costs. This is a new type of system design for Sweden but is found in other countries in the world. At Bo01 in Malmö (new housing area built for a housing fair) the solar collectors are connected directly to the district heating network. When the solar collectors produce more heat than is required within the area, the surplus heat can be distributed in Malmö's district heating network. The disadvantage is that the working temperature in the solar collectors is affected by the temperature in the district heating network. Normally this is a relatively high temperature level, which lowers the efficiency of the solar collectors but reduces the investment costs*

30–40%, which can be of interest for new, well-insulated buildings. With the higher heat output, the output per m² of solar collector will be lower.

It is normal for the solar heating to be designed to cover the summer load within the district heating area. Then a short-term store is used for solar heat. This type of large-scale solar heating project has a market today, not least within the EU. Projects with diurnal stores and a given heating load during the summer period are simple to design and implement, and the results and the profitability follow given patterns.

7.3 Seasonal heat stores

Solar heating plants with seasonal heat stores are normally designed to cover up to 70–80% of the annual heating load. In practice this means that the annual heat coverage ranges between 60% and 90% depending on variations in climate. The solar collector area is 2 m² per MWh annual heat demand, and about 2–3 m³ storage volume of water (or the equivalent) is needed per m² of

solar collector area for colder climate conditions. When a solar heating system with a seasonal heat store is undersized, the volume needed for storage is reduced. This results in lower specific investment costs per m² of solar collector, but at the expense of a lower solar fraction.

Solar heating systems with seasonal storage have a considerably higher potential solar fraction than other system solutions. Therefore Swedish R&D resources were granted early on to develop a technically and economically viable seasonal storage technology for solar heating systems.

One of the first larger solar heating projects in Sweden was at Ingelstad outside Växjö (see Figure 2.14). Solar collectors (1325 m²) were connected to a seasonal store of 5000 m³ of water. The solar heating was designed to cover the main part of the space heating demand in a housing area comprising 52 single-family houses.

Research on seasonally stored solar heat continued during the 1980s and 1990s, and several projects have been started in Sweden following the Ingelstad project.

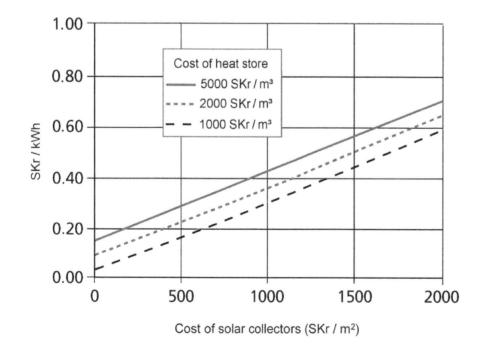

FIGURE 7.3 *Heat costs with a 10% solar fraction. The diagram shows the cost of systems with short-term stores as a function of the cost of the solar collectors and the cost of the stores*
Source: Dalenbäck and Åsblad (1994)
9 Skr = approximately €1

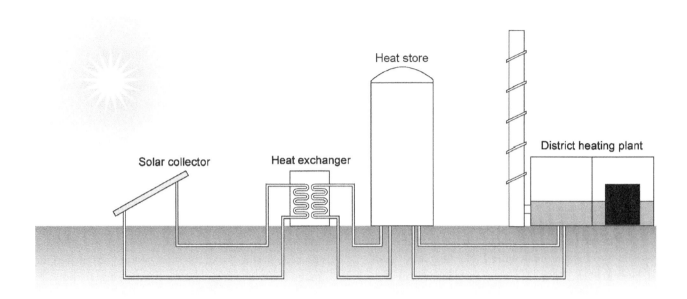

FIGURE 7.4 *Most of the large-scale solar heating plants that have been built in Sweden store heat in a short-term store (diurnal store). The size of the heat store is determined in relation to the solar collector area, and is usually 0.1 m³ storage volume per m² of solar collector. The tank is normally welded on site, and should be well insulated and preferably of a good height to maximize temperature stratification. The advantages of a short-term store are that it is cost-efficient and that the volume can be used as a standby, for example if the auxiliary boilers need service*

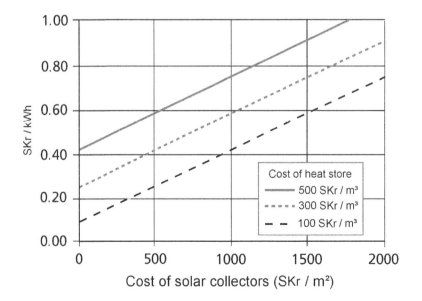

FIGURE 7.5 *Heating costs with a 70% solar fraction. The diagram shows the cost of systems with a seasonal store as a function of the cost of the solar collectors and the cost of the heat store*
Source: Dalenbäck and Åsblad (1994)
9 Skr = approximately €1

The most well known is the Lyckebo project outside Uppsala, where solar heat is seasonally stored in a large rock cavern that contains 100,000 m³ of water. The heat store was sized for 20,000 m² of solar collectors but there are only 4300 m² there today, so the water is also heated by electricity (which can be off-peak electricity). With a seasonal store the solar fraction can be up to 80% of the total heat load. Approximately 2–3 m³ of heat store is required per m² of solar collector. There is a far greater potential for storing solar heat with seasonal storage than with conventional systems. The rock cavern store can be considered of interest if, for example, existing oil reserve stores (or the knowledge gained from constructing these) can be used.

At the end of the 1980s there was a pilot study of the possibilities for seasonally stored solar heat in Kungälv (Claesson *et al.*, 1988). The aim was to investigate the possibility of using solar heating for a built-up area of 350,000 m² floor area, with a total heat load of approximately 56 GWh/year. The study was based on a 75% coverage by solar heat, to be accomplished with about 126,000 m² of solar collectors and a rock cavern store of almost 400,000 m³. At 1988 prices the total heating cost for the production and distribution of solar and other auxiliary heat would be about 0.42 SKr/kWh (€0.05/kWh). The proposals in the study were never implemented (Figure 7.6). In Malung and Särö (outside Göteborg, Sweden) there are projects that are prototypes for housing areas that get part of their heat supply from seasonally stored solar heat.

7.4 Key ratios

There is a relationship between the cost of solar heating and the choice of heat store. The total construction cost is the sum of the cost of the solar heating system and the storage costs. The actual heat output related to the capital cost (the annual interest and repayment costs) gives the cost of the heat from the solar heating plant. The heat cost can be calculated as a function of the given heat output together with the cost of the solar

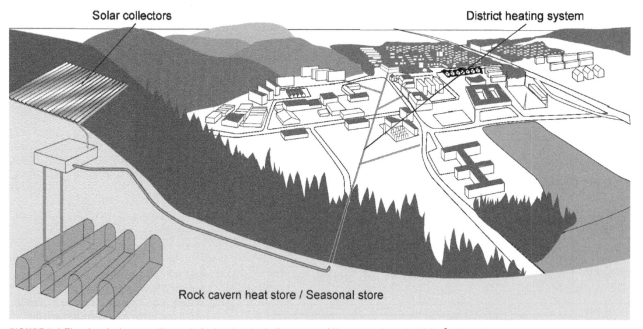

FIGURE 7.6 *The sketch shows a pilot study for heating the built-up area of Kungälv, using 126,000 m² solar collectors and a rock cavern store of almost 400,000 m³, to give 75% solar coverage. However, the proposals in the pilot study were never carried out*
Source: Claesson et al. (1988)

collectors and store. Even if the investment cost per unit volume is large for a short-term store it affects the cost of solar heating relatively little. The opposite is true for seasonal storage, where the investment cost per unit volume is relatively small, but affects the solar heat cost to a very great extent.

Large-scale solar heat projects are all individual, and so it is difficult to generalize regarding solar collector area, suitable storage volume or total heating costs. Furthermore it is not known today what will be achieved by the development of technology and rational mass production. However, the cost of large-scale solar heating projects is undoubtedly being reduced, and this is increasing their competitiveness.

Table 7.1 lists examples of plant size, and Table 7.2 examples of sizing and cost.

Table 7.1 Examples of plant size

Heat load (GWh/year)	System	Solar collector area (m²)	Storage volume (m³)
2	Short-term store (tank)	650	65
	Seasonal store (tank, pit store)	4300	12,000
10	Short-term store (tank)	2500	250
	Seasonal store (tank, pit store)	21,000	50,000
60	Short-term store (tank)	19,000	1900
	Seasonal store (rock cavern)	120,000	390,000

Source: Dalenback and Åsblad (1994)

Today solar heating plants for local heating plants and district heating systems can be built in Sweden for 1200–1500 SKr (€133–167) per m² of solar collector where the production of energy from the solar collectors approaches 400 kWh/m² annually. The heat cost in the latest large-scale solar heating project in Sweden is 0.35–0.4 SKr/kWh (approximately €0.04/kWh) (without subsidies) with 20 years' depreciation period at a real interest rate of 6%.

Table 7.2 Examples of sizing and cost

System type Heat load	Today's technology SKr/kWh (€/kWh)		Advanced technology SKr/kWh (€/kWh) technology/marketing/store	
Short-term store 10%				
2 GWh/year	0.76	(0.08)	0.623/0.38/0.365	(0.07/0.04/0.04)
10 GWh/year	0.60	(0.07)	0.510/0.305/0.295	(0.06/0.04/0.03)
60 GWh/year	0.45	(0.05)	0.365/0.230/0.226	(0.04/0.03/0.03)
Seasonal store 70%				
2 GWh/year	1.025	(0.11)	0.91/0.705/0.61	(0.10/0.08/0.07)
10 GWh/year	0.73	(0.08)	0.64/0.50/0.445	(0.07/0.06/0.05)
60 GWh/year	0.53	(0.06)	0.46/0.375/0.34	(0.05/0.04/0.04)

The cost of solar heating is calculated with an annuity factor of 0.10.
Source: Dalenback and Åsblad (1994)

7.5 Operation and maintenance

Running and maintenance costs for large-scale solar heating projects are low. It is impossible to claim ownership of solar radiation. Therefore the flow of energy from the sun can be regarded as free and is difficult to tax. The running costs for pumps and other equipment are very low. A typical sum for the total running and maintenance costs of large-scale projects in Sweden is 0.03 SKr (<€0.01) per kWh produced or 5% of the total investment cost. There are projects where the running and maintenance costs have been as low as between 0.5% and 1% of the total investment cost.

The Swedish Council for Building Research report R104:1988 (Claesson et al, 1988) on seasonal storage of solar heating in Kungälv is one of the most comprehensive reports on large scale solar technology that has been published in Sweden. Maintenance costs for the project comprising 126,000 m² solar collectors and a seasonal water store of 400,000 m³ were calculated as a percentage of the construction costs for the whole project not just the solar heating system (see Table 7.3). Maintenance costs are calculated as a percentage of the construction costs for the whole

Table 7.3 Annual maintenance costs (as a percentage of construction costs)

Boiler plant	2%
Building	0.5%
District heating system	1%

Source: Claesson et al. (1988)

project, not just for the solar heating system. Staffing is estimated as three person-years/year.

The direct running and maintenance costs for solar heat can be estimated as 3–5% of the total construction cost. The Monitoring Centre for Energy Research at Chalmers University of Technology has compiled actual results for running and maintenance costs for solar heating projects in Falkenberg and Nykvarn.

7.6 Summary

If solar heating is to make a real breakthrough as an alternative form of heating and an important part of our global heating supply, development of large-scale solar heating technology is essential. Large-scale solar heating plants make the technology acceptable. Where these plants are constructed, know-how is spread, giving inspiration and often stimulating further solar heating investments. In connection with this type of project, the manufacture of the solar collectors can also be improved, rationalized and industrialized. Large-scale solar heating projects also allow advanced analysis and monitoring to provide a basis for optimization of the technology.

There are two main types of system: those with short-term stores and those that use some form of seasonal storage. The latter still need to be developed to reach a commercial breakthrough. Systems with short-term storage generally attain an annual solar fraction of 5–15%.

Large-scale solar heating plants are primarily intended to be used in district heating plants and for DHW production for hospitals, housing areas or other large consumers of DHW and heating. The market for this sort of plant is therefore extremely large.

They can be designed to cover up to 100% of the heating load, but this would be unreasonably expensive with solar fractions of over 80–85%. When designing and sizing it is the required amount of heat from the sun that is of interest, not – as is the case with other heat sources – the heating power that is to be achieved. One of the positive contributions of solar heating is that a higher system efficiency can be achieved. By replacing the low efficiency of low-load combustion with solar heating, the efficiency of the whole system is improved.

Solar heating plants on the ground require relatively large areas. To avoid overshadowing, and to make maintenance of the solar collectors easier, approximately 3 m² of ground area is required for every m² of solar collector, which obviously imposes certain limitations. The question of available land must be solved at an early stage. A combination of locations is possible, where part of the area of solar collectors is placed on the roof and part on the ground.

In solar heating plants on the ground, solar collectors with a loadbearing framework can be used. This means that most flat-plate solar collector structures and evacuated solar collectors can be used in this type of project. Site-built solar collectors placed on the ground have been constructed with the absorbers in complete lengths.

■ 8. Economy and profitability analysis

Solar energy has a unique advantage as the flow of energy is free. For other heating alternatives there is always uncertainty in the development of energy prices. The profitability of a solar heating plant is determined mainly by capital costs, together with changes in the cost of energy. Therefore interest and depreciation costs are decisive in the profitability analysis; solar heating has negligible running and maintenance costs. Solar heating investments also give a number of synergy effects and coordination gains, for example the use of a common storage tank, and these should be taken into account in the overall evaluation. When making economical comparisons it is also essential to be consistent in reasoning between the different alternatives, for example by comparing capital costs and fuel costs in the same way.

8.1 Methods of calculation

Profitability analysis can be based on many different conditions and several different calculation models. The total heat cost for the solar heating plant is composed of capital and running costs. A common method for larger solar heating plants is to calculate the capital cost according to the annual instalment method. The capital cost can be calculated using an annuity factor. The interest rates and the depreciation period determine the annuity factor. The investment is multiplied by the annuity factor to give the annual capital cost, which – together with the running cost – is divided by the solar

heat production to obtain the cost of the solar heating per kWh:

$$\text{cost/kWh} = \frac{(\text{annuity factor} \times \text{investment}) + \text{running}}{\text{solar heat production}}$$

This can be done by assuming a depreciation period of 25 years and a real interest rate of 6%, which would then correspond to an annuity factor of 0.0782. In this type of calculation the running costs can be estimated as a percentage of the investment, for example 1–3%.

Box 8.1 Fixed price for 25 years!

With the annual instalment method capital costs are determined by the user, who can decide the interest rates and the depreciation period. On condition that the interest rates are reasonably stable during the life of the solar heating plant there is a fixed price for solar energy during the calculation period, which in some cases can be as much as 25 years. There are not many other types of energy that can offer this stability in price!

Another way of calculating profitability is to compare the investment of the solar heating plant with the value of the energy/heat that it replaces. This method calculates how many years it will take for the plant to pay for itself. A simplified way is to use the value of electrical heating and exclude interest costs. In this case the investment of the solar heating plant is divided by the value of the electrical heating that is saved annually, and this gives the repayment period:

$$\text{repayment period} = \frac{\text{investment of solar heating plant}}{\text{value of energy replaced}}$$

Finally, solar heating costs can be calculated using an equation in which the plant's total capital cost (interest + depreciation costs) and the estimated running costs are divided by the number of kWh that the plant produces, calculated yearly. To carry out the calculation the following information should be obtained:

- total investment
- estimated running and maintenance costs
- period of depreciation
- interest costs during the period of depreciation
- amount of produced/replaced heat annually.

The form of calculation is as follows. Let investment (own capital) × interest rate (interest rate on own capital − tax effect) = interest costs; let depreciation cost per year = depreciation; and let heat production in kWh/m² per year = heat production. Then:

Year 1:

$$\text{cost/kWh} = \frac{\text{interest costs + depreciation}}{\text{heat production}}$$

Year 10:

$$\text{cost/kWh} = \frac{\text{interest costs (−10 years' depreciation)} + \text{depreciation}}{\text{heat production}}$$

Year 20:

$$\text{cost/kWh} = \frac{\text{interest costs (−20 years' depreciation)} + \text{depreciation}}{\text{heat production}}$$

8.2 Practical example: profitability analysis

A simple calculation sheet for profitability analysis of solar heating for outdoor pools is presented in Box 8.3. The calculation sheet can of course be used for other solar heating investments. It works very well as the basis for any sort of application.

Box 8.2 How to determine the interest rate and depreciation period

The rate of interest used in the calculation has a decisive impact on the profitability of a new solar project. Interest costs can be calculated in different ways. Normally the interest rates in a profitability analysis are based on the interest rate on a loan. The interest rate can also be related to the possible alternative returns on the capital that is tied up, for example a bank's interest on deposit. The difference between the interest rate on a loan and the interest rate on a bank account can amount to several percentage units. This can have a significant impact on the calculation.

As the length of life of a solar collector is very long (up to 25–35 years) a realistic depreciation period for solar heating investments is 20–25 years. As the calculation of profitability is based mainly on the capital cost, the calculation period determines how profitable the investment will be. The long life is therefore an important economic factor.

Box 8.3 Profitability analysis for pool facility in Sweden

Pool and present heating

1	Pool area	m²
2	Total energy requirement	kWh/year
	(area × heat demand, 700 kWh/m²/year)	
3	Energy price	€/kWh
4	heating cost	€/year
	(energy price × heat demand)	

Solar heating installation

5	Solar collector area	m²
6	Solar heating production/year	kWh/year
	(200–400 kWh/m² absorber area)	

Profitability analysis

7	Savings	€/year
	(Energy price × solar heating production/year)	
8	Installation cost	€
	(according to tender or rough estimate with €/m² × absorber area)	
9	Coordination gains	€
	(no need for further investment due to solar heating installation etc.)	
10	Repayment period	years
	(net investment divided by electricity savings)	

Annual cost

Interest cost	€/year
Amortization payments	€/year
Auxiliary heating	€/year
Running/maintenance costs	€/year
Total yearly costs	**€/year**

Source: Sten-Ivan Bylund, Bygginfo

■ 9. Procurement

9.1 Survey

Careful analysis of the prerequisites is required for a successful solar heating investment. An early and thorough survey of the physical basis is essential. This includes the different locations for the solar collectors, space for the storage tank and the system equipment, and the pipe runs to and from the solar collectors. It is also important to note and consider the advantages of combinations of systems where solar heating can make a contribution, and the synergies that can be gained. Of course the will to invest and the costs for alternatives must also be established. See the checklist in Box 9.1.

9.2 Methods of procurement

Forms of procurement differ from one country to another, as there can be considerable cultural and legislative differences. This book is based on present Swedish conditions.

There are two main methods of procurement for larger solar heating projects. In a *general contract* complete documents are prepared by a building services consultant on the commission of the owner. It is an advantage if the consultant has documented experience of solar heating systems. Building contractors submit tenders for the work, and the solar heating plant is then constructed by the successful contractor, following the documents prepared by the consultant. In some cases the owner divides the contract into several contracts with the aim of influencing the choice of contractor for each trade. A divided contract assumes an active owner who has the knowledge and ability to influence both the design of the plant and the total cost.

In *a design and construct contract* the owner has prepared an overall description that is a basis for procurement. The overall description gives the requirements for the solar heating plant and the terms to be included in the contract. To allow fair tendering it is important that the overall description is clearly formulated and that all essential parameters are given, so that the tender procedure will be correct and on equal terms. The plant is then constructed by a selected contractor, who is responsible for the work and the main part of the design. The main contractor engages the consultants and subcontractors that he considers necessary to carry out the work. In a design and construct contract the contractor himself is responsible for the operation of the plant according to the overall description.

One way to reduce the design costs for this form of contract is to divide the design into several parts. For example, the overall description for the solar heating plant can be agreed as a system description together with energy and temperature requirements. The detailed design of pipes, pumps, storage tank etc. can be delegated to the general contractor. If the contract is divided up, be careful to give clear demarcation regarding the extent, content and guarantees.

Box 9.1 Start the right way

Make a survey of the prerequisites; evaluate suitability
Are the basic conditions suitable for solar heating? This is the main determinant for investment. Check the summer load; check that roof areas are available, and that there is space for pipe runs and the storage tank. Apart from this the other decisive factor is the possibility of designing a heating system to which solar collectors can be connected. The will to invest and the profitability requirement are other essential factors.

Design and size accurately
The solar heating plant will give the best results (for both heat production and economy) if the size is optimized. The area of the solar collectors and the volume of the storage tank must be suited to existing conditions. Correct input data is essential for good preliminary studies and design and sizing work.

Choose a technique with quality
A solar collector is exposed to tough external conditions – everything from snow, ice and wind to very high working temperatures (in some cases over 250°C). Choose a solar collector that has been tested by a certified testing station. The testing stations have lists of solar collectors that have passed the initial tests, thereby giving a form of quality label.

Orientate the solar collectors correctly
Choose a favourable orientation (see Table 2.1), and avoid overshadowing. Solar collectors produce no heat at all when completely in shadow.

Heat store
The storage tank is often the most important component in a heating system and therefore affects the operation, output, economy and, in some cases, the convenience (service and need for inspection). The design and construction of the heat store determine the efficiency of the system. Decisive factors are, for example, temperature stratification and heat losses (degree of insulation) and also that the overall system is suitable for the heat producers included and correctly connected, according to the supplier's instructions.

Choice of components
All component parts in the solar collector loop are exposed to the heat transfer fluid, which has specific properties and high working temperatures. To prolong the operation and service life of the system, the components must be of the quality to stand the stresses they are subjected to. Some components must be inspected annually, and this must be made clear in the running and maintenance manual (for example the freezing point and corrosiveness of the heat transfer fluid, the function of the pump and control unit, cleaning of the filter, solar collector fixings). The regulations and recommendations for the relevant country must also be followed.

Control devices
Some safety components are obligatory in the system, such as the safety valve, and the manometer for reading pressure. Other components are advisable, to be able to check the operation: for example temperature sensors on the supply to and return from the solar collectors and at different levels in the storage tank. A flowmeter is also useful. A time recorder on the pump for the solar loop is useful to be able to compare operation times between different periods (the same month, different years and so on).

Choose the supplier carefully
Choose a supplier who can give satisfactory guarantees, references and running and maintenance manuals. It is a great advantage if the supplier has a local representative (retailer) with regard to the possibility of quick service and inspection. In the same way the stock kept of the most important spare parts and the quality of the products delivered should be checked.

Visit a plant in operation
To get an idea of the technology a visit should be made to a plant in operation. The user can give valuable information on operation from experience.

9.2.1 SPECIFICATION OF REQUIREMENTS

Regardless of the form of procurement, the design and sizing should be based on the actual heat demand and be presented together with the calculation methods used. The proposed system design should be presented together with the specification and outline drawings. It is particularly important to specify outline drawings, temperatures and energy/heat demand in a general contract.

A common reason for lower output than expected from the solar collectors is a lower heat consumption than that given in the design specification. For this reason it is important that the assumptions and measurements for heat demand are reasonable and correct.

The specification of requirements should state that the solar collectors shall be tested and preferably certified by an authorized testing station. In all circumstances calculations should state the heat output from the solar collectors, and a recommendation from an accredited testing station, a university or equivalent body should preferably be included.

To be able to make an overall evaluation of the tender these documents must also give instructions on running and maintenance and on the relevant guarantee conditions. This must be stated in the complete documents (overall description). Even in the tender documents there must be a clear description on the follow-up of performance, how this is to be verified, and manuals for annual inspection and maintenance. It is difficult or impossible to make complaints in retrospect if these are not specified at the tender stage.

The extent of the guarantees varies with the type of solar heating system, the supplier and the form of procurement. In some cases it is only a guarantee for materials, but, for example, in a general contract guarantee conditions can be considered that include the heat output of the solar collectors (kWh/m^2 of solar collector per year). In a design and construct contract it is possible to go a stage further by agreeing on the total heat output of the plant – that is, the annual saving in oil or similar. In this case it is important that the parties agree on the form of control, the division of responsibility, and the compensation to be paid on departure from the agreement.

In evaluating the tenders there are many guiding parameters. They should be compared based on a comprehensive evaluation of earlier references (prospects of carrying out their commitments), given costs and the accompanying documents (basis for calculation, descriptions, drawings and guarantees). The evaluation of the solar collectors should be based on the specific cost for the respective makes, where the solar collector's cost per m^2 is compared with the certified heat output, assuming similar systems and sizing. To evaluate the tender documents in a design and construct contract it may be necessary to enlist the help of a building services consultant with previous experience of solar heating.

For larger solar heating projects an independent supervisor should check the construction of the plant so that it is carried out in agreement with the tender documents. The completed plant should be inspected by a inspector with previous documented experience of solar heating systems. A guarantee inspection should be carried out after two years. For a guarantee of performance the inspector should approve this based on the signed contract.

The simplest form of performance control is the installation of a heat meter to record the heat output of the solar collectors. Note that in larger plants, with computerized operation monitoring, the performance control can be integrated for a marginal extra cost. However, in smaller systems such test methods entail considerable extra costs.

Box 9.2 sets out a basic checklist for purchasing.

Box 9.2 Purchasing

Stage 1: Design and sizing
▨ Demand
▨ Requirements
▨ Consequences

Stage 2: Specification of requirements
▨ Space, possible positions and other external factors
▨ Flexibility demands
▨ Energy dependence
▨ Service costs
▨ System demands, required combinations
▨ Environmental impact

Stage 3: Profitability analysis
▨ Willingness to invest
▨ Payback time/amortization period
▨ Interest charge
▨ Profitability requirement
▨ Running costs (convenience)
▨ Price stability (availability)

Stage 4: Form of contract
▨ Guarantee conditions
▨ Service requirements
▨ Follow-up/energy guarantee

■ 10. **Other solar technologies**

E ven if water-based solar heating dominates the world market there are other forms of application, which are attracting increasing interest. Solar electricity is one form of application that has always met with a great deal of interest, and which is now expanding strongly both in Europe and in the rest of the world. Passive solar technology and solar air heating are also discussed in this chapter.

10.1 **Solar electricity**

Solar electricity is actually produced in all power plants except nuclear ones (though solar energy is nuclear hydrogen fusion energy, in the sun!). Fossil fuels and biofuels are chemically stored solar energy in plants (photosynthesis), and the enormous solar energy flows in the atmosphere cause wind and rain, which can be converted into wind and hydropower.

Only the direct conversion of solar radiation into electricity will be dealt with here. This can be carried out by solar cell or PV systems (photovoltaic systems) or by conventional power plants using steam from concentrating high-temperature solar collectors. The latter is also called *thermal solar electricity*. In the USA several larger solar thermal power stations were built during the 1980s, of which the most recent has a power of 80 MW. There is about 400 MW of thermal solar electricity in California alone. Thermal solar electricity is approaching profitability even for grid-connected applications in large systems located in sunny areas.

PV systems are still fully competitive only for stand-alone applications. A stand-alone PV system competes in cost with grid extensions, which may cost €10,000–20,000/km. The cost for a PV system can be lower for a small load, even a few hundred metres from the grid. In a city with paved roads the distance can be much shorter, only tens of metres. The market for distributed grid-connected systems is also increasing rapidly now because of innovative subsidy systems in some countries. Today the accumulated total installed PV system capacity is over 1000 MW_p, and during 2001 the market was as large as 350 MW_p. The growth rate is also very large, in the range of 20–30% per year.

Experience from larger PV plants in Europe shows small or negligible running and maintenance costs. Almost no moving parts are needed in the system. The most common problem is a fault in the inverter. The computer control system may also require some maintenance, but otherwise solar cell technology is very reliable.

Solar cells generate electricity directly from solar radiation and can be used in all sizes of system, even mobile ones (on boats, caravans and electric cars, for example). The conversion of sunlight to electricity takes place without any substance being used up. The electricity production is completely without environmental impact. Many people are convinced that solar cell technology is the solution for our future energy supply. In a future scenario, with electrically driven vehicles, it will also be possible to provide for the transport sector with electricity produced, to a large extent, by solar cells.

For many people the potential of solar cell technology as an electricity source is still unknown. This is mainly because the technology has been underrated, but also because the costs have been too high. There is now some more direct investment in the development and manufacture of solar cells and system components technology. The major development work is taking place in the USA and Japan, but Europe has begun to show her paces.

With new materials and with research in new production methods, including thin-film technology, the price of solar cells will be drastically reduced. With better components (regulators, inverters and system accumulators) at a lower cost, the outlook for the solar electricity market will also be more favourable for grid-connected systems. Today, the system costs make up as large a part of the whole system as the solar cell costs. As prices decrease and new applications become possible, large expansion of the market can be expected within the building sector. A project in Germany aimed at constructing 100,000 solar cell roofs is an example of this trend. New ways will be found for integrating solar cell technology into building roofs and facades.

10.1.1 HISTORY

Solar electricity is not a new phenomenon. It was discovered as early as 1839 by the French scientist Edmund Becquerel (1820–91). In his studies of the solar spectrum and phosphorescence he discovered what came to be known as the *Becquerel effect*. Becquerel's discovery of the transformation of solar radiation directly into electricity was the starting point for solar cell technology. The technology was ignored for most of the 19th century owing to the lack of knowledge of the real energy content of sunlight. The Swede Knut Ångström's accurate measuring instrument constructed at the beginning of the 20th century constitutes a milestone here.

Practically applied solar cell technology did not really get started until the 1950s, when semiconductor

technology became available. Also, the need for a lightweight and reliable energy source for satellites and remote telecommunications stations gave enough resources for research and industrialization of the silicon solar cell at Bell Laboratories.

The acceleration increased, with a high end-use for telecommunications both on earth (telecoms masts) and in space (satellites), keeping pace with the introduction of electronics (and then principally the rapid development of semiconductor technology) during the second half of the 20th century. New electronic developments leading to very energy-efficient products have also resulted in more PV applications.

Until now the main material resource for solar cells has been silicon 'scrap' from the semiconductor industry. However, in future, continuing market growth will be very dependent on the production of new, low-cost, solar-grade silicon.

10.1.2 TECHNOLOGY

Solar radiation can be seen as a flow of light particles called *photons*. The energy of the photons is inversely proportional to the wavelength of the light. Solar light is a spectrum of photon energy, from low-energy light (infrared) to high-energy light (ultraviolet). Light that is absorbed in the solar cell transfers the light energy to electrons. A small part (5–30%) of the incident solar energy can be converted into electricity that can be used in an external electric circuit. The larger part is converted into heat in the solar cell.

Solar cells, whose construction in most cases is based on crystalline silicon, consist of a thin wafer (or layer) of semiconductor material with contacts on the front and back. When the solar radiation impinges on the solar cell, the cell is polarized so that the front becomes negatively charged and the back positively charged (Figure 10.2). The metal contacts on the front and back take up the charge in the form of an electric current. The cell efficiency and the intensity of the sunlight determine the amount of electricity that the solar cell produces. The silicon solar cell works with up to 20% efficiency, which corresponds to 200 W/m^2 in clear sunshine in Sweden. As the voltage from a single

FIGURE 10.1 *Solar cells with reflectors, Carissa Plains, California, USA. The power output of the solar cell plant is 16.5 MW. By allowing the panels to follow the sun and placing reflectors on the support, the annual output is doubled compared with that of conventional plants*
Photo: Lars Stolt, Uppsala University, Sweden

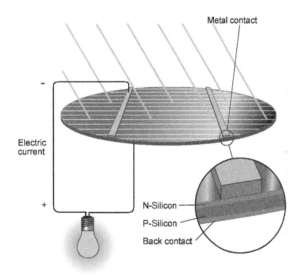

FIGURE 10.2 *A solar cell that is exposed to light is polarized so that the front is negatively charged and the back positively charged. Sunlight is transformed into electricity with the help of semiconductors, materials that are less efficient conductors of current than metals, but are not insulators. Normally silicon is used as the semiconductor material. Metal contacts on both sides collect the charge, which can be used as current in an external circuit. A silicon solar cell gives approximately 0.5 V, which is too low a voltage to be useful in practice. Therefore a number of cells (30–36) are coupled in series to reach a total suitable voltage – to charge a 12 V lead–acid battery, for example*

silicon cell is only 0.5 V, several cells are connected in series. Usually a module voltage suitable for charging 12 V lead–acid batteries (such as car batteries) is chosen. This normally requires 36 silicon solar cells connected in series in a module.

The new polycrystalline or multicrystalline solar cells are considerably cheaper than monocrystalline silicon solar cells. Today's research and technology development is aimed mainly at reducing the cost, both of the PV modules and of the system components. The lifetime is already proven to be at least 25 years for silicon technology. The reliability of a well-designed system is also proven to be very high.

Swedish researchers are focusing their efforts on developing thin-film technology. The goal is to maintain high efficiency at an acceptable production cost and energy consumption during manufacture (at present it takes, in the worst case, approximately 5–10 years' operation time to produce the amount of energy used in the manufacture of a crystalline silicon solar cell).

The new thin-film technology is very interesting because of the minimal use of material and the possibility of rationalized manufacturing methods.

Thin-film technology using amorphous silicon (which has a high energy absorptance) is established on the market. The PV module efficiency is low (4–6%), and there is some doubt about the quality and whether the cells will be able to compete with other materials in the future. Cells made of several thin layers stacked on top of one another are beginning to appear on the market, and have about 10% efficiency.

Table 10.1 Efficiency in theory and in practice

Material	Bandgap (eV)	Theoretical efficiency (%)	Laboratory cell (%)	Solar cell module (%)
Silicon	1.1	29	23	20
Gallium arsenide	1.4	31	25	
Thin film material				
Amorphous silicon	1.7	27	12	9
Copper indium diselenide	1.0	27	15.4	11
Copper indium gallium selenide	1.2	29	16.9	
Cadmium telluride	1.4	31	15.8	10

Note: where no values are given for solar cell modules there is no data available due to little or no manufacture.
Source: Johnson (1993)

Another thin-film solar cell is the *Grätzel cell*, named after its inventor Michael Grätzel (Swiss Federal Institute of Technology, Lausanne, Switzerland). The Grätzel cell is an electrochemical cell that is partly composed of electrolyte fluid. The semiconductor material is titanium dioxide (TiO_2), and light absorption takes place in a dye. The advantages of the Grätzel cell are its relatively simple manufacturing method (and hence low production costs), and the ready availability of titanium oxide. Tests are being carried out on consumer goods such as watches.

The annual electricity production from solar cells located on the available roof area is sufficient for the household electricity for a family in a private house (see worked example in Box 10.1). One problem is that the electricity demand does not completely coincide in time with the available sunshine. In a stand-alone

---

Box 10.1 Solar electricity for single-family house: worked example for Sweden

In a single-family house of 100 m² with a normal roof slope of 45° and a good orientation to the south, the available south-facing roof area is 70 m². The roof is fitted with solar cells with an efficiency of 12%, which means that 120 W/m² is received with full solar radiation (1150 kWh/m² per year). The rated output of the plant is 8.4 kW peak power. The annual electricity production of the solar cells will be

12% × 70 m² × 1150 kWh/m² × 0.85 = 8211 kWh

(The factor 0.85 is to take account of various small losses.)

---

system, therefore, some sort of electricity storage is needed.

The usual way to store solar electricity is to use electric accumulators, such as batteries. The size of the accumulator that is needed depends on the consumption and the requirements for availability and storage capacity margin. A stand-alone solar electricity plant consists of the solar cell modules, a battery bank, and regulators for charging and discharging (see Figure 10.3). A solar electricity system with battery accumulators is a d.c. system and often of low voltage. It can be supplied with an inverter for the relevant a.c. grid voltage and frequency if desired.

In a grid-connected solar cell system the storage problem is solved by using the electricity grid as a store. When the solar production is larger than the local consumption, PV energy is delivered to the grid, and other power plants in the grid can reduce their production marginally. As seen from the point of view of the grid and the power plant it is equivalent to turning off a large load locally.

To speed up the breakthrough of solar cell technology, the system components must be improved and become cheaper. There is also a need for investment in research and development on the system side, a part that makes up about half the cost of a solar electricity plant today.

FIGURE 10.3 *Where there is no grid-connected electricity a solar cell plant is often the cheapest alternative. Common areas of use in Scandinavia are electricity supply for holiday homes, telecommunications stations in the mountains, and other equipment with low power demand remote from the ordinary electricity grid. Complete standard packages can be bought. The solar cell system then includes solar cell panels, battery (accumulator), charging units and generator* Photo: Neste Advanced Power Systems AB, Stockholm, Sweden

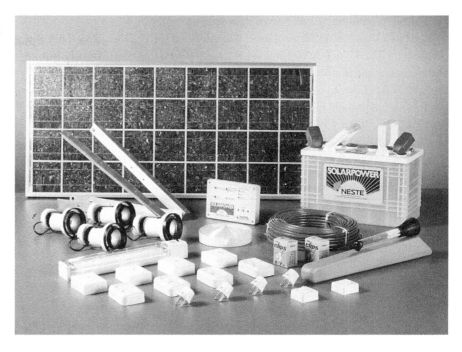

10.1.3 AREAS OF USE

The interest in solar cell technology and applications increased during the oil crisis of the early 1970s. One early niche was the so-called *space cells.* Nowadays all new satellites get their power from the sun. Remote telecommunications stations and masts are often PV powered too. Solar cells are now used in many consumer products such as watches, pocket calculators, radios, mobile phones and battery chargers. The solar cell market has expanded in pace with the development of the cells and the technology, and now also includes radio stations, lighthouses, the electrification of isolated buildings and villages, and telecommunications stations. A clear niche for solar cells is in places situated far from the electricity grid, where it is more cost-effective to use solar cells than grid-connected electricity. In stand-alone applications, PV systems also replace diesel- or petrol-driven electric generators, which are noisy, polluting and expensive, both in maintenance and in operation.

During the 1990s the solar cell industry was established in earnest. World production in 2001 had a power output of about 350 MW$_p$ and corresponded to an area of solar cells comparable to more than 100 football fields. Of the total production, stand-alone systems are still the most important application. There has been considerable interest in two growth markets. One is in developing countries, where pilot plants are being built, for example for refrigerating medicines, telecommunications and water purification and pumping plants alongside more traditional lighting projects. The other growth area is in grid-connected large-scale power plants and distributed grid-connected PV systems integrated in buildings. An example of this is shown in Figure 10.4.

The solar cell market has an annual growth rate of about 20–30% (Figure 11.4). In 2001 total sales turnover was about €3500 million, and the area of solar cells sold was almost 1.5 km². The PV market is an emerging industry sector that already employs tens of thousands of persons all over the world. The modularity of PV systems also means that the market can reach out into decentralized areas and create welfare, new jobs and income for local people in developing countries.

FIGURE 10.4 *The Österäng project, AB Kristianstadsbyggen (Sweden), was built in 1998 with two parallel systems with a solar cell area of 100 m². The designed power output of 59 monocrystalline solar cells is 2 × 6.5 kW and they are expected to produce 11,000 kWh per year*
Photo: Per Sandell, Neste Advanced Power Systems AB, Stockholm, Sweden

Table 10.2 shows a typical breakdown of costs for a solar cell plant in a larger project. The constant development of solar cells and other system components will probably lead to changes in investment costs and the breakdown of costs. Accumulators (batteries) and inverters are considered to have great potential for being both better and much cheaper. A clear trend can be seen. Grid-connected plants increased their share of the market and reached 68% of the total market in 2001. The general reduction in price seems to be continuing: statistics for 2001 show that prices were reduced by 4%. Also new, more energy-efficient appliances play a very important role in reducing system costs, as the PV electricity user most often buys comfort

(such as light, cooking, ventilation, refrigeration, telecoms, the Internet, computer use or water pumping) and not kWh.

Table 10.2 Typical breakdown of costs for a solar cell plant in a larger project (>1 MW)

Cost item	Type of solar cell module	
	Crystalline silicon	Thin film
Modules	55%	40%
Frame	10%	20%
Inverter	13%	15%
Computer control	4%	4%
Cost of design and installation	10%	12%
Other	8%	9%

A realistic vision for the future in Sweden is for the production of 5 TWh of solar electricity per year. This amount does not require a large storage capacity in the grid. A production of 5 TWh requires about 6 GW$_p$ of solar cells, which need a total area of 60 km², or 7 m² per inhabitant.

Worldwide, solar electricity can have an important function in the grid as a distributed peak power electricity source for air conditioning loads in warm and humid climates, as for example in the southern USA. In stand-alone PV applications the possibility of solar home or village PV systems is one of the most

Box 10.2 Ringön shopping centre

Stockholm Energi has installed one of Sweden's largest solar cell plants on Ringön shopping centre. A total solar cell area of 76 m² produces electrical energy that is fed into the electricity supply in the building. The plant consists of 90 monocrystalline silicon solar cells, delivered by the Swedish solar cell company GPV (Gällivare Photovoltaics).

The efficiency is approximately 13%, and at full solar radiation the plant delivers a total of 9900 W power (35 V and slightly over 3 A). Four inverters transform the direct current to 230 V alternating current. The efficiency in transformation to alternating current is approximately 90%. During a normal 'solar year' (availability of solar radiation) the plant is estimated to generate between 7000 and 8000 kWh of solar electricity.

interesting, as it can both create local jobs for installation and maintenance and deliver energy to new local industry and thus create new jobs and income. In this stand-alone situation solar thermal systems can also be used, of course, to increase the standard, comfort and health situation.

Looking further into the future, solar cell technology is expected to compete with conventional electricity production in the grid. But this is competition against a movable target, and the timeframe is very unclear. Meanwhile, PV systems already have many cost-effective niche applications that will support further development and industrialization.

Box 10.3 Solar cells: some facts and figures

- ▨ To produce 1 TWh electricity/year in Sweden requires an area of 10 km² solar cells, which corresponds to 1m² of solar cells per inhabitant.
- ▨ If the roof of a parking place were equipped with solar cells, electricity required for running an electric car approximately 3000–5000 km would be produced annually.
- ▨ Solar cells with an efficiency of approximately 15% and a total area of 5425 km², the size of Lake Vänern (the third largest lake in Europe), would cover the whole of Sweden's electricity consumption (146.5 TWh including losses in 2000)!

10.1.4 DESIGN AND SIZING: WORKED EXAMPLES

The design and detailed calculation of solar cell plants should be carried out by experts. However, it can be useful to know the key ratios, numbers and methods of calculation, and these are indicated in the worked example below.

Example 1: Large stand-alone PV system

1   First determine the consumer's average energy demand per day (24 hrs), Q_d, given in kWh/day. (Also consider reduction of the demand by more energy-efficient appliances.)

2   Determine the average insolation on the solar cell, H_m, in kWh/m² per day for the normal operating period or season of the system

3   Determine the F factor that corrects for losses between the module and the user in the solar cell system. Typical values of F are between 0.7 and 0.8.

4   I_{max} is the radiation nominal solar intensity during testing of solar cell modules = 1.0 kW/m².

5   The total required nominal peak power of the solar cell panel in kW, P_{tot}, can now be determined by the equation

$$P_{tot} = \frac{Q_d \times I_{max}}{H_m \times F}$$

where P_{tot} is the total nominal (rated) power of the solar cell panel in kW; Q_d is the average energy consumption per day (24 hrs) in kWh; I_{max} is the solar radiation nominal intensity during testing of solar cell modules in kW/m² (at present 1.0 kW/m²); H_m is the average daily insolation on the solar cell in kWh/m² per day; and F is a factor that corrects for system losses. Typical values of F are between 0.7 and 0.8.

6   Assuming a total load, Q_d, of 7.25 kWh/day, an average solar radiation on the solar cell of 5.9 kWh/m² per day and an efficiency factor, F, equal to 0.8, the total installed nominal PV power needed can be calculated as follows:

$$P_{tot} = \frac{7.25 \times 1.0}{5.9 \times 0.8}$$

The calculation shows that 1.5 kW nominal peak power is suitable. With present-day panel power output this corresponds to a total PV module area of 12.5 m².

7   The size of the batteries should be chosen to cover about four days' electricity requirements, without charging from solar radiation. This gives a total storage requirement of

7.25 kWh × 4 days = 29 kWh

A 75 Ah 12 V battery can store up to 1 kWh (75 Ah × 12V/1000 = 0.9 kWh) of electrical energy.

Therefore, as a rough estimate, about 30 batteries of this size will be required. Here, for this large system size, an expert should be consulted to get a more precise battery type and size in the individual case. Normally the batteries are not designed for 100% discharge, in order to extend their life. A maximum 50% discharge level will greatly increase the life of the battery.

Example 2: Small stand-alone PV system

As a further worked example, consider the electricity requirement for a summer cottage, with only small d.c. loads, that is used at the weekends.

1   The solar cell panel is estimated to produce 75% (corresponding to the F factor above) of the total nominal peak power of 60 W, which is 45 W.

2   The estimated electricity requirement of the summer cottage (see Table 10.3) is estimated as 154 Wh/day. In ampere hours (Ah) this will be

$$\frac{154\,\text{Wh}}{12\,\text{V}}$$

The total consumption will be 13 Ah per day, and 26 Ah over the weekend.

Table 10.3 Estimated electricity requirements of a summer cottage in Scandinavia

Use	Hours per day	Power (W)	Wh/day
Water pump	0.2	35	7
Reading light	3.0	5	15
Working light	2.0	13	26
Fan	4.0	1.5	6
TV	2.0	50	100
Total			154

3   This amount has to be produced during the remaining 5 days of the week. For an estimated average of 2 hours' sunshine per day, during the charging time of 5 days, this means 10 hours of sunshine. The consumption during the weekend will be:

$$154\,\text{Wh} \times 2 = 308\,\text{Wh}$$

If this is to be replaced during 10 hours' sunshine, the minimum requirement will be 31 W (= 308Wh/10h). But the solar cells must be larger because of the previously mentioned loss. They should therefore have a power output of 41 W (= 31W/0.75).

4   To cover the weekend's electricity requirements the battery accumulator may not lose more than 50% of its capacity. Therefore for weekend requirements the battery should be at least 26/0.5 = 52 Ah.

5   A suitable plant must also be designed to cover 10 days' electricity consumption without being charged during this time. Therefore the battery capacity must be at least 10 × 13 Ah, so in this case the recommended battery size is 130 Ah.

6   To avoid greatly reduced output, all forms of shading must be avoided. A position free from overshadowing is very important in the summer as well as in the winter.

10.1.5 PRACTICAL HINTS FOR INSTALLING PV SYSTEMS

■   The performance of the PV module is very sensitive to shading. Locate the modules as high up as possible and away from high trees and other shading objects. This is much more important than the optimum tilt and orientation.

■   It is very important to use corrosion-resistant electrical materials, preferably of the same metals, in all electrical connections. This will prevent corrosion, and added resistance and voltage drops, over the years. This is in fact one of the most common failure modes in small PV systems. The PV modules almost never fail.

■   The batteries can also be very reliable in small systems, but they can also cause severe problems and hazards for the inexperienced. A modern system controller greatly improves the life of the battery, as it prevents both overcharge and deep discharge and mixes the acid regularly by short gassing periods.

■   Battery safety aspects also have to be considered. Lead–acid batteries produce hydrogen and oxygen in a perfect explosive mixture during overcharging, and are then very dangerous. Good ventilation and a good controller are very important. Lead–acid batteries, even in small sizes, are also very powerful, and a main electrical fuse in the system close to the positive battery connection is recommended.

■ Avoid both extremely low and especially high battery temperatures by placing the batteries indoors if possible. Typically a 10° higher average battery temperature will halve the lifetime because of faster internal corrosion.

■ Always use highly efficient appliances in the system. They are almost always more cost-effective than a larger PV area.

Consider the possibility of using solar thermal collectors and other heat sources instead of PV modules for some loads. For example, a dishwasher or washing machine can be bought with separate hot and cold water inlets, so that the water does not have to be heated by electricity. One kWh of solar thermal heat costs only about one tenth of one kWh of PV electricity! This, of course, also applies to water heating for other purposes.

10.2 Passive solar heating

In a passive solar heating system there are no moving or mechanical parts i.e. the building is placed and designed so that it makes use of the heat provided by the sun. (In an active solar heating system, solar energy is converted using active components.)

The use of solar energy as a passive heat source is no new phenomenon. The Ancient Greek philosopher Socrates referred to passive solar technology. In Socrates' solar house the indoor climate was improved by the door openings being placed so that the sun could reach far into the house and thus warm up the greater part of the building structure. The accumulated heat was given off during the dark and cold hours. During the summer, when the sun was high in the sky, a projecting roof cut off the sun's rays, with the aim of preventing excess temperatures indoors. Even the native North Americans are said to have utilized passive solar heat to

improve the indoor climate. There are many historical examples around the world to learn from.

Passive solar energy covers a number of different areas. The first that springs to mind makes use of the incident solar insolation by the position and choice of windows (energy-efficient technology). The contribution that can be made to passive solar buildings by the admittance of daylight should not be underestimated. It reduces the need for energy for electric lighting and affects the energy balance in the building, as the excess heat from lighting will be reduced. Passive solar technology also includes the use of the thermal mass of the building, which will store and distribute the heat over the day and night. The distribution of heat in passive solar buildings is dependent on the plan of the building, and heating may be improved by pre-heating the fresh air supply (see section 10.3.1). To a certain extent the building's location in its surroundings can also be considered as making use of passive solar energy (that is, taking into account the topography, shading, areas of water, wind, vegetation etc.).

Many countries in the warm areas (around the Mediterranean, for example) utilize solar heating for ventilation in buildings and for evening out the temperature between day and night. Traditional building elements (windows, walls, roofs, and so on) are designed in such a way that solar radiation is used improve the indoor climate. In the Scandinavian countries, *low-energy buildings* have been developed that are extremely well insulated, with small window areas facing north and the larger windows orientated to the south: the goal is to minimize the auxiliary heat.

The principle of a passive solar heating system is very simple. The building is constructed to admit solar radiation. Building materials are chosen that are able to store and later emit (distribute) the stored heat. The heat storage is diurnal. Seasonal storage is not possible with present-day technology. To increase the heat input it is important that the sun's rays are allowed to penetrate as far into the building as possible, to distribute the heat to as large a part of the building as possible. Heat stored in heavy materials (such as stone,

FIGURE 10.5 *The basis for utilization of passive solar heating is that the building structure allows solar energy to be absorbed and stored in heavy building materials (the thermal mass), e.g. concrete floors with a stone, brick or tile finish and/or concrete, brick or stone walls. For maximum utilization the solar radiation must be allowed to penetrate far into the building. By drawing out the eaves a shadow is created that cuts off the sun's rays when the sun is high in the sky. In this way the risk of excess temperature in the building is reduced during the period of most intensive solar radiation. In some cases it may even be necessary to ventilate out the excess heat*

brick or concrete) gives the best indoor comfort (Figure 10.5).

In passive solar heating, the heat from the incident solar radiation is stored directly in the structure of the building (Figure 10.6). It is an advantage if the building has a heavy structure that can optimize the storage function. The aim is for the building structure to use solar heating to even out heat between the day and the night, and for ventilation purposes by improving the natural ventilation. Solar energy can be stored in a stone, brick or concrete wall and/or in the floor during the day. At night the walls and/or the floor emit (give back) the heat that has been stored. Heavy materials (the thermal mass) can store large amounts of heat without their surface temperatures increasing and thus causing discomfort. The heavy building materials act as a heat buffer, which spreads the stored heat over the day and night. Optimization of the indoor climate places great demands on the control system of the normal heating supply, but a certain amount of indoor temperature variation must be accepted.

Buildings with large thermal mass (heavy structure) have a slow system: that is, heat transmission takes place slowly (see Figure 10.7). In extreme cases, conditions are created for a cooler indoor climate during the summer months and warmer during the winter, compared with buildings constructed of light building materials. A typical example is provided by stone churches, which often feel cool during the summer.

Passive solar energy does not require any mechanical assistance; however, the orientation and building construction must be designed to optimize insolation

Box 10.4 Energy-efficient windows

Windows are important components in passive solar heating systems. It is necessary to create a balance between unwanted cooling and unwanted heat. Nowadays there are windows that are well insulated and use different techniques to avoid unnecessary heat losses.

and storage capacity. The choice of the quality and performance of the windows is also of great importance.

Buildings that use passive solar energy are not common in Europe at present, but these theories may be found in some architectural circles. Plans for passive solar heating are well advanced in the EU: the goal is to

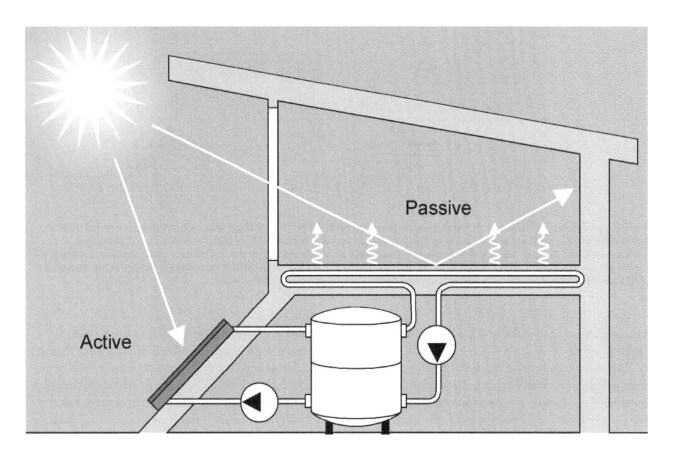

FIGURE 10.6 *Passive solar heating utilizes solar radiation without mechanical help, as opposed to active solar heating, which uses a pump to transport the heat transfer fluid in the solar loop*

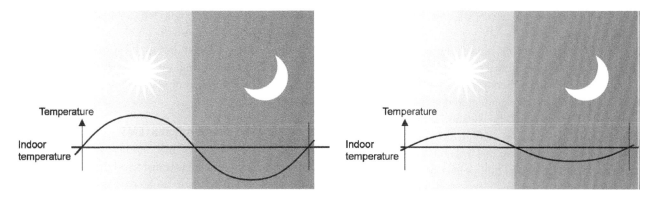

FIGURE 10.7 *Heat can be stored in heavy materials without too great a rise in the surface temperature. In this way solar energy from the daytime radiation can be stored, to be given off during the remaining hours of the day. If there is no heavy material in the building there is an increased risk of extensive solar radiation quickly giving rise to unwanted heat in the room. The variation in temperature over 24 hours will be relatively large (left). If, however, there is a balance between radiation (sun access) and the thermal mass (the proportion of absorbent heavy material) the variation in temperature in the room is evened out (right)*

reach the equivalent of 400 TWh passive solar heating before 2010, which will probably increase interest in the technology!

Passive solar heating systems have many advantages. They do not require any mechanical parts, which cuts down the investment costs. The incident solar insolation can be utilized without running and maintenance costs, and the system lasts as long as the building.

10.3 Solar air heating

There are two main types of active solar heating system: one that uses water as the heat transfer fluid, and one that uses air. As solar heating is usually used for heating water, the water-based systems are dominant. However, interest in air-based systems is increasing, partly because they are often simpler in construction, and the investments costs are therefore lower.

In a solar air system the solar radiation is converted into heat, and air is used as the heat transfer fluid. Hot air is normally transported mechanically using some sort of fan, but there are also natural convection systems. Heat can either be stored in some part of the building structure (part of the wall, for example) or used to pre-heat the ventilation air.

Solar air collectors are constructed in approximately the same way as conventional water-based collectors. The construction can, however, be simplified because air is used as the heat transfer fluid. Solar air heaters are therefore generally cheaper to construct per unit area than water-based ones. It is also easier to fit them into the building envelope, and it is not unusual for solar air collectors to be integrated in the building structure, either as roof modules or as the outer skin of the wall. In the latter case part of the facade of the building is a transparent layer backed by an absorber. In some cases the intention is to pre-heat the ventilation air for the building where the supply air passes through some sort of solar collector. This can be either a simple construction (a cover glass with an absorber behind) or a glazed sun space/conservatory where the air supply is warmed up before passing into the building. The temperature in the sun space is allowed to vary throughout the year, and can be higher or lower than the acceptable indoor temperature. The sun space can be a heat source for a heat pump. It also acts as a buffer zone that gives protection from the weather and in this respect fills an important heat-saving function. To cover 10–20% of the annual heat demand for heating ventilation air, 10 m² of solar air collectors are required per dwelling, if the solar collector gives approximately

150 kWh/m² in an average year (figures for the Scandinavian climate).

The advantages of solar air heating are that the system is simple and can make use of low working temperatures. Solar air heating is cost-efficient because the solar collectors can be integrated in the building envelope. The system is also cheaper, as certain parts of the building structure are used for storing and distribution of heat. In general, solar air systems also have an advantage in that they deliver heat quickly. They can be used in several ways: space heating during the spring, autumn and winter can be combined with DHW production in the summer. The risk of corrosion and leakage is much less in a solar air system than in a water-based system.

One of the disadvantages is that air is a less efficient heat transfer medium than water and is more difficult to regulate and distribute. This means that larger ducts must be used to distribute the heat. It is also important to avoid or minimize the sound made by the fans in the system, which can be transmitted through the air ducts in the building. The electricity demand of the fans in the system must be minimized. Where open systems are used, in which the pre-heated ventilation air is fed directly into the indoor environment, it must be ensured that this does not cause problems of dust or damp.

10.3.1 OPEN SOLAR AIR SYSTEMS

In open solar air systems the ventilation air supply is pre-heated in a solar collector without any form of storage. Heat transfer is fast, and the heat is led quickly into the building. The size of the systems varies from small systems for holiday cottages (see Figure 10.9) to larger systems for industrial use or multi-family dwellings. A cost-efficient type uses unglazed perforated solar collectors, in which the air is forced through many small holes in a metal absorber. Figure 10.8 is a sketch of a simple solar-heated drier for hay and cereals. There are approximately 2,000,000 m² of solar air collectors in use to dry cereals around the world, everything from simple constructions to more sophisticated ones of the type shown in the figure. Figure 10.9 shows a simple air collector, which has

become increasingly popular when the main requirement is to increase the air changes and give marginal additional heat during the spring and the autumn. The target group for this kind of system is mostly summer cottages or single-family houses in warm climates.

10.3.2 CLOSED SOLAR AIR SYSTEMS

In these systems air is led in a closed circuit through the solar collector, and heat is transferred to some sort of heat store. The heat store can be part of the building, for example part of the wall structure, but the heat can also be stored in a traditional water store. In these cases it is normal for the solar air heater to give additional heat to the ventilation system during the cold part of the year; the excess heat during the summer can be transferred to a hot water tank and be used for DHW. There are also system solutions in which the heat is stored in pebble beds next to the building. Closed solar air systems are normally integrated in the building, which makes them relatively cost-efficient.

FIGURE 10.8 *A number of solar-heated hay and cereal driers were installed in Sweden at the beginning of the 1980s. A simple construction was based on a system of ducts in the roofing material, and the heated air was forced into a drying room with the help of fans. During the early 1980s there were subsidies for this type of installation*

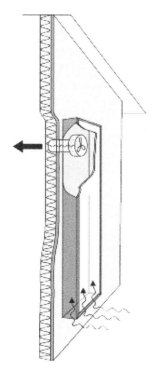

FIGURE 10.9 *This simple solar air collector has become increasingly popular in Sweden. A simple box structure is placed vertically on the external wall. Fresh air is heated in the box and a fan (preferably driven by a solar cell) forces the heated air into the house. A large target group for this type of system is the owners of holiday cottages who want to have a little extra heat during the spring and autumn, and increase the ventilation at the same time. The vertical position means that the summer output is limited*

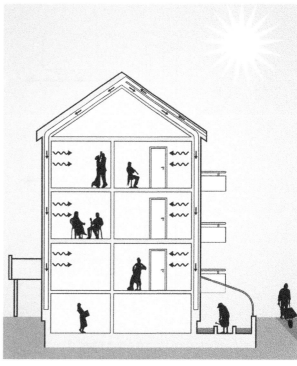

FIGURE 10.10 *The architect Christer Nordström (Askim) has installed 350 m² of air solar collectors on a multi-family dwelling in Göteborg. The solar collectors produce more than 100,000 kWh per year, of which slightly over 70% is used for space heating*
Source: Christer Nordström Arkitektkontor AB, Askim, Sweden

Christer Nordström, an architect in Göteborg in Sweden, has been successful in his work with large-scale solar air heating projects, in and outside Sweden. In Järnbrott (see Figure 10.10) and Gårdsten housing areas (outside Göteborg, Sweden) multi-family dwellings have had solar air collectors installed. The projects have aroused a great deal of interest.

10.3.3 SOLAR WALLS

In a solar wall the building facade is clad with a transparent layer that lets the solar radiation pass through and at the same time protects the wall from damp. There are different types of solar wall (which can be either open or closed systems). Some store the heat in the wall, and it is later given off into the building. Others are ventilated solar walls using thin metal sheets, which absorb solar radiation, placed behind the transparent layer. The heated air can be fed into the building from an air space between the metal sheet and the wall. A new building technique has been developed allowing the construction of a type of solar wall. New transparent insulation materials allow the solar radiation to pass through at the same time as they are an insulating material which thereby reduces the heat losses from the building.

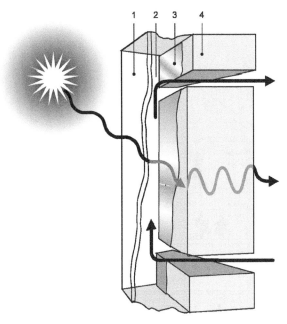

FIGURE 10.11 *A trombe wall consists of a transparent outer layer (1) separated by an air space (2) from a heavy wall (4), which has an absorbent surface (3) (e.g. painted black or covered with a foil). Solar energy passes through the transparent layer, is absorbed by the heavy wall, and passes slowly through the wall to the room behind. The air in the room can also be heated directly by passing it through the air space if this is connected to the room, but the vents must be closed at night to prevent the heated air from being cooled in the air space. Trombe walls and similar solar wall structures can be of interest both technically and economically in connection with facade renovation of existing buildings*

10.3.4 SOLAR CHIMNEYS

In natural ventilation systems ventilation can be increased by the exhaust air from the building being heated. The thermal forces can be strengthened by the exhaust air being heated by a special type of solar air heater – a *solar chimney* (Figure 10.12). The aim of solar chimneys is not to give extra heat to the building but to raise the temperature of the exhaust air over that of the ambient air. In this way the natural driving forces are strengthened (called the *stack effect*).

FIGURE 10.12 *Tångaskolan in Falkenberg, Sweden: a natural ventilation system with solar chimneys has been installed when refurbishing. (architect: Christer Nordström) Photo: Lars Andrén*

◼ 11. **Summary and a look to the future**

Recent years' progress in research and development in solar energy technology has meant an increased competitiveness within both the solar heating field and solar cell technology. The market for both of these products is expanding strongly today, not least in Europe.

The trend in solar heating technology is towards increased interest in large-scale projects. More and more systems are being built for connection to district and local heating plants in Europe. Solar heating technology for multi-family dwellings is also being standardized, and this will probably give good competitive strength in the future. Even within the single-family house market the trend is for solar heating technology to be integrated in the traditional system technology. By standardizing solar heating technology the investment costs are reduced; at the same time it is easier to sell systems as solar collectors become a natural part of the heating or DHW system.

It is a question of utilizing solar heating based on the existing demands and the available solar insolation. The system must be designed for these conditions, and can therefore differ from region to region. Solar heating installation is easier if there is a tradition of using central heating systems and if the building standard is normally high. It is also an advantage if there is also an expressed environmental goal and if investments with long depreciation periods are customary.

Investment in solar heating can be considered an energy-saving measure where the aim is to reduce the need for purchased energy. Roof-integrated solar collectors are a good example of this application. It is also possible to regard solar heating investment as a heat production unit for heating of, for example, local heating plants or swimming pool facilities.

To speed up the market development of solar heating, a new approach is needed among building services consultants, architects, builders, tradesmen, and managers. Solar collectors must be regarded as a natural part of the building. It is also important that planned refurbishing and extension programmes do not limit the opportunities for solar heating applications.

The solar heating technology of the future is obvious within the housing sector. There will be an increased use, mainly in new buildings but also for refurbishing and repairs. Owners of houses or properties that face a necessary replacement of the heating system should see solar heating as an obvious alternative. The integration of solar collectors in the heating system must be natural for those who design and sell heating systems in the future.

Better-developed heat storage technology would encourage the use of solar heating in the future. A heat storage technology in which some of the summer surplus could be stored for use in winter would speed up the development.

Researchers at Vattenfall's Älvkarleby Laboratory in Sweden have already increased heat production from solar collectors by about 30% with the help of reflectors, a technique that can be of interest for larger

Table 11.1 Large-scale heating projects in Europe. The seven largest are based completely on Swedish technology

	Plant/Year	Owner/Country	Area m²
1	Kungälv, 2000-	Kungäv Energi AB, **SE**	10,000
	Marstal, 1996-	Marstal Fjernvarme, **DK**	9,043
	Nykvarn, 1984-	Telge Energi AB, SE	7,500
	Falkenberg, 1989-	Falkenberg Energi AB, SE	5,500
	Neckarsulm, 1997-	Stadtwerke Neckarsulm, **DE**	5,044
	Ærøskøping, 1998-	Ærøskøping Fjernvarme, DK	4,900
	Rise, 2001-	Rise Fjernvarme, DK	4,000
	Ry, 1988-	Ry Fjernvarme A/S, DK	3,040
	Hamburg; 1996-	Hamburger Gaswerke, DE	3,000
	10 2MW, 2002-	ENECO Energy, **NL**	2,900
	Friedrichshafen, 1996-	Techn. Werke Friedrichsh., DE	2,700
	Nordby, 2002-	Samsø Energiselskab, DK	2,500
	Groningen, 1985-	De Huismeester, NL	2,400
	Breda, 1997-	Van Melle, NL	2,400
	Anneberg, 2002-	HSB Brt Anneberg, SE	2,400
	Augsburg, 1998-	Bayerisches Staatsministerium, DE	2,000
	Fränsta, 1999-	Vattenfall AB, SE	1,650
	Stuttg.-Burgholzhof, 1998-	Neckarwerke Stuttgart AG, DE	1,635
	AS Stadion, 2002-	nahwaerme.at GmbH & Co KG, **AT**	1,407
20	Bo01, 2001-	Sydkraft Värme Syd AB, SE	1,400
	Hannover-Kronberg, 2000-	Avacon AG, DE	1,350
	Ekoviikki, 2002-	**FIN**	1,258
	Säter, 1992-	Säter Energi AB, SE	1,250
	Eibiswald, 1997-	Nahwärmegen. Eibiswald, AT	1,250
	Lisse, 1995-	Dames&Werkhoven, NL	1,200
	Älta, 1997-	Vattenfall AB, SE	1,200
	Bad Mitterndorf, 1997-	Genossensch.. Biosolar BM, AT	1,120
	Neuchatel, 1997-	Swiss Fed Office of Stat., **CH**	1,120
	Kockum Fritid, 2002-	Sydkraft Värme Syd AB, SE	1,100
30	Innsbruck, 1999-	Wohnen am Lobach, AT	1,080
	Salzburg, 2000-	Gem. Salzburger Wohn. m.b.H., AT	1,056
	Åsa, 1985-	EKSTA Bostads AB, SE	1,030
	Tubberupvænge, 1991-	Herlev kom. Boligselskab, DK	1,030
	Saltum, 1988-	Saltum Fjarnvarme A/S, DK	1,005
	Odensbacken, 1991-	Örebro Energi, SE	1,000
	Fjärås Vetevägen, 1991-	EKSTA Bostads AB, SE	1,000
	Stuttgart-Brenzstr., 1997-	Neckarwerke Stuttgart AG, DE	1,000
	Rostock, B-höhe, 2000-	WIRO mbH, DE	1,000
	Hågaby, 1998-	Uppsalahem AB, SE	930
40	Kullavik 4, 1987-	EKSTA Bostads AB, SE	920
	Poysbrunn, 1997-	Genossensch.. B/SW Poysbrunn, AT	870
	Hammarkullen, 1985-	Gbg Bostads AB, SE	850
	Göttingen, 1993-	Stadtwerke Göttingen, DE	850
	Ekerö, 1997-	Ekeröbostäder AB, SE	800
	Nikitsch, 1997-	FWG Nikitsch, AT	780
	Brandaris, 1999-	Patrimonium WS Amsterdam, NL	760
	Kroatisch-Minihof; 1997-	FWG Kroatisch-Minihof; AT	756
	Särö, 1989-	EKSTA Bostads AB, SE	740
	De Zwoer, 1990-	Stichting Zwembad Dr.-Rijsenburg, NL	740
50	Nordhausen, 1999-	SK GmbH Nordhausen, DE	717
	Gårdsten, 2000-	Bostads AB Gårdsten, SE	705
	Oederan, 1994-	SWG Oederan mbH, DE	700
	Lienz, 2001-	Stadtvärme Lienz GmbH, AT	690
	Echirolles, 1999-	OPAC 38, **FR**	689
	Henån, 1997-	Orust kommun, SE	685
	Magdeburg, 1996-	Universität Magdeburg, DE	657
	Malung, 1987-	Malungsbostäder, SE	640
	Älta, 1998-	HSB Brf Stensö, SE	600
	Kungälv, 2001-	Kungälv Energi AB, SE	600
60	Obermarkersdort, 1995-	Fernw.genossench. Oberm., AT	567
	Ottrupgaard, 1995-	Ottrupgaards bofaellessk., DK	565
	Chemnitz, 1998-	Solaris Verwaltungs GmbH, DE	540
	Heemstede, 1998-	Stichting De Hartekamp, NL	520
	Steinfurt-Borghorst, 1999-	W & T Bau GbR, DE	510
	Torsåker, 1999-	Vattenfall AB, SE	500

Source: Jan-Olof Dalenbäck, Chalmers University of Technology, Göteborg, Sweden

solar collector fields or where large-module solar collectors are placed on flat roofs . See Figure 11.1.

FIGURE 11.1 *Alby project, near Stockholm, Sweden: heat production from the solar collectors is increased by approximately 30% by placing reflectors between the rows of solar collectors Photo: Göran Bolin, Solsam Sunergy AB, Stockholm, Sweden*

In the long-term perspective it is probable that much of the world's energy supply will be based on solar energy, partly for reasons of resources but above all for environmental reasons. Thus the potential is very great, much greater than can be imagined. It is not unlikely that worldwide there will be some hundreds of TWh of solar heating. It should be possible to use the greater part of this energy to replace the use of oil and electricity for heating purposes through new combination plants based on solar energy and biofuels. Heating systems with both solar and biofuels are suitable for single-family houses as well as multi-family dwellings and smaller local heating plants (district heating systems).

A considerable expansion of solar heating can be seen in Europe today. Increased interest and better system technology, together with regional and national market support, resulted in an annual installation of over 1,000,000 m² solar collectors per year at the beginning of the 1990s, and this has subsequently grown by almost 20% per year during the decade. Figure 11.3 shows how the market for glazed solar collectors more than doubled during the 1990s.

Table 11.2 Amount of installed solar heating in Europe

Country	Accumulated installed m² 1975–2001	Share of Europe market (%)	Installed 2000 (m²)	Market growth(%)
Austria	1,791,000	16	150,000	10
Belgium	26,900	3	3400	18
Denmark	332,000	0	25,000	0
Finland	26,000	0	7000	0
France	329,700	3	15,000	25
Germany	3,805,000	33	615,000	46
Greece	2,995,000	27	170,000	6
Ireland	2100	0	400	25
Italy	307,000	3	28,000	25
Netherlands	204,000	2	27,500	11
Norway	9500	0	2000	25
Portugal	231,000	2	5500	9
Spain	409,000	4	41,000	34
Sweden	195,200	2	18,200	10
Switzerland	295,000	3	27,000	0
UK	168,000	2	10,000	

It is noticeable that the rate of growth has been strong in recent years. For the whole continent the growth rate was approximately 40% during 2000: Germany showed the greatest increase, and has about 60% of the total installed solar collector area in Europe today. For 2001 it is estimated that about 1,493,700 m² solar collectors were installed, which shows clearly that the rate of growth has slowed down. Everything points to the rate of growth having decreased even more during 2002. Stagnation can be seen in certain dominant countries (e. g. Germany and Denmark), above all due to energy and subsidy policies
Source: ASTIG (2002)

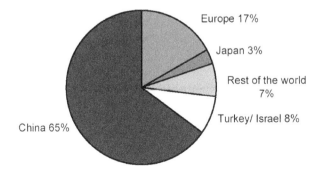

FIGURE 11.2 *It is estimated that over 8,500,000 m² of solar collectors were installed during 2001. By far the largest market is China. The statistics are not completely reliable, however, but the amount installed is estimated as 5,500,000 m² during 2001. During the same period almost 1,500,000 m² of solar collectors were installed in Europe; the North American market is estimated as being 120,000 m²*
Source: ASTIG (2002)

However, this rate of development has stagnated during the first year of the 21st century. Germany and Austria are two of the largest markets. Spain has a rapidly growing solar heating market; the Spanish state of Catalonia has recently decided that 60% of DHW supply should be covered by solar heating in all new buildings.

Solar technology in all its forms will be a natural feature in the architecture of the future and in future heating systems. Solar heating is an excellent producer of DHW for sports facilities, camp sites and other facilities with large and seasonal hot water consumption.

However technology development proceeds, whatever the consequences of the environmental debate, and irrespective of the changes in energy prices solar energy will be of considerable importance in our energy supply. Above all the carbon dioxide situation in the world will force a development of technology that reduces the environmental impact.

The potential for solar electricity is, if possible, even more difficult to foresee, but it is far greater than for solar heating. When solar cells are competitive, there will be an explosive expansion. However, this requires an increase in the price of grid electricity or a corresponding reduction in the price of solar electricity. To generate 1 TWh of electrical energy about 5 m² of solar cells are needed per single-family house, and this corresponds to 7 million m² of solar cells.

Solar-based electricity production is inevitable following changes in the world's use of energy, as industry becomes increasingly electricity-intensive, and as concern for the environment grows, with the problem of the greenhouse effect and fears about atomic power, as well as the aim of achieving a more sustainable society. The world's environmental situation and the supply of raw materials will decide the expansion of solar electricity.

The rate of development of solar cells is rapid. For many years (particularly during the 1970s) the USA had a leading role in the development of the technology. Japan also increased her efforts during the 1990s and is now making great progress (Figure 11.5). Japan's state

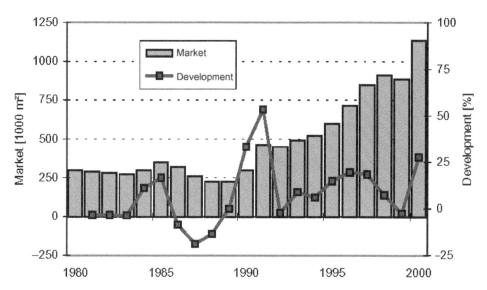

FIGURE 11.3 *The market for glazed solar collectors has more than doubled during the last decade. During 2000 more than 1,000,000 m² of glazed solar collectors were sold in Europe*
Statistics compiled by: Jan-Olof Dalenbäck, Chalmers University of Technology, Göteborg, Sweden
Sources: 1980–1985 from Sun in Action; 1996 is the differences between the European Solar Industry Federation (ESIF) totals in 1996 and Sun in Action; 1997–1998 mainly from Observer; 1999 mainly from Deutscher Fachverband Solarenergie, Gleisdorf, Sept 2000; 2000 mainly from Deutscher Fachverband Solarenergie

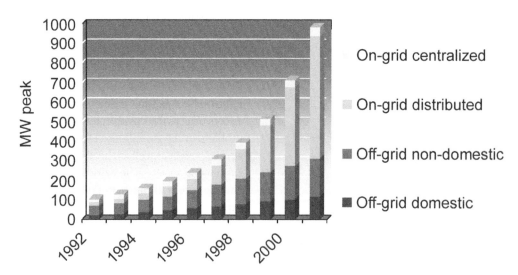

FIGURE 11.4 *Cumulative installed PV power by application areas in IEA-PVPS countries (MWp)*
Source: IEA-PVPS (2002)

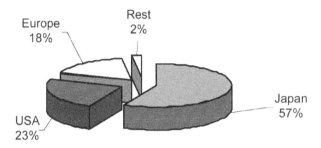

FIGURE 11.5 *Japan still has a dominant share of the world market in solar cell module production. Manufacture, research and even the use of solar cells are completely dominated by the industrialized world, although it is the developing countries that have the greatest need of the technology*
Source: IEA PVPS (2002)

programme for the installation of 80,000 solar cell roofs has made the country one of the world's foremost manufacturers of solar cells. However, since the middle of the 1990s Europe seems to have become the driving force. The countries that have contributed to this include Germany, Italy, France, the UK, Spain and Switzerland. As well as the great increase of production in Europe, large investments in systems are also being made. In Italy 18 MW of installed power was reached by the end of 1999. As a step towards this, Europe's largest solar cell power plant was built outside Naples

(3.3 MW). Another interesting example is that Switzerland had a goal of 50 MW of grid-connected solar electricity for 2000, but reached only 15 MW during that year. Germany is investing 4.4 thousand million SKr (€490 million) to equip 100,000 roofs with solar cells that will together deliver 350 MW to the electricity grid. The UK has recently decided to invest €30 million for the installation of solar panels around the country.

Figure 11.6 shows the various sectors' shares of the solar electricity market in 1996 and 2001.

Sweden's work during the latter part of the 1990s, and in particular the results achieved by the Ångström Solar Centre at Uppsala University, have attracted attention. Research there is concentrated on CIS solar cells and Grätzel cells, and at times they have held the world record for efficiency of the CIS thin-film solar cell.

Everything points to thin-film technology being the most interesting. The CIS solar cell shows good long-term stability and high efficiency, and the manufacturing method has very promising characteristics, including high utilization of material. The Grätzel solar cell is in an earlier stage of development but has great development potential. The use of reflectors to increase irradiation is also an

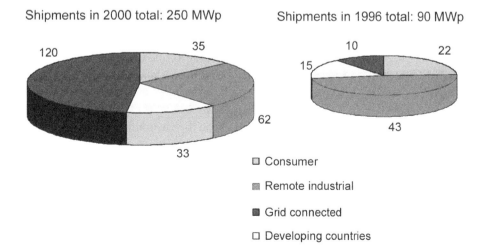

FIGURE 11.6 *The figure compares the division into market sectors and shows how they have changed between 1996 and 2000*
Source: P Maycock

interesting technique.

A development of hybrid systems can also be considered in regions with heating needs, where electricity and heat production are combined in the same unit. There are already solar cell prototypes that can generate heat by cooling the hot surface of the solar cell with a stream of liquid or gas.

The commercialization of grid-connected electricity generation with solar cells requires a great reduction in price. With a reduction equivalent to a factor of 10 (which is a realistic expectation in the long term), solar cell technology will be profitable as long as it is not necessary to store the electricity. For example, in Sweden this would mean that 5 TWh solar electricity could be produced for 0.3–0.4 SKr/kWh (€0.03–0.04/kWh). This can be compared with the current estimated energy cost, for a 100 kW plant, of 5.05 SKr/kWh (€0.56/kWh) (see Box 11.1).

The next stage in development is electricity production for peak demands where accessible solar energy and electricity consumption correspond: plants where the solar cells can simply cover the peak load when integrated in a building or as necessary in an inefficient electricity grid (far from the production source). It is not until after this that accumulating systems or alternative use, for example for hydrogen production or use in the transport sector, will be of interest.

A large future potential for solar cell technology is expected to be in the form of grid-connected plants in which solar cell modules are integrated in the roofs and facades of buildings (see Figure 11.7). A grid-connected solar cell plant transforms direct current from solar cells to alternating current with an inverter, to be fed into the normal electricity grid and/or to be used in the building. When there is surplus production, electricity is sold *to* the grid, and when the solar electricity is not sufficient, electricity is bought *from* the grid in the normal way.

There will be an increased use of solar electricity, following the progress that is being made in solar electricity technology. Solar electricity can be competitive today where electricity from the grid involves large investment costs, for example where

---

Box 11.1 Solar electricity: current energy cost per kWh

For a 100 kW plant the following energy cost (2000) per kWh (excluding design, installation and possible subsidy) can be calculated:

Total investment costs:	5.8 million SKr (€644,444)
Annual production:	100,000 kWh
Interest:	6%
Payback time:	20 years
Annual capital cost (instalment):	505,000 SKr (€56,111)
Energy cost per kWh:	5.05 SKr (€0.56)

---

there are few connections (consumers) in a geographically widespread area. The electricity supply for holiday cottages is a good example. Great investments are being made in Europe, the USA and Japan on both the research and the manufacturing sides. Large investments are also being made by the users. This leads to falling costs at the same time as systematic research increases the efficiency. The average efficiency of a solar cell is slightly over 15% today. The long life and the high second-hand value improves the financial prospects for solar cells, and will probably speed up the market breakthrough.

The great breakthrough for solar energy will probably not come until solar electricity is cost-competitive. Then a substantial part of the world's energy supply can be from solar cells. Even today a development in this direction can be seen.

There can be no doubt of the future importance of solar energy for the energy supply of the world. We shall be forced to make ever greater use of the enormous energy flows of the sun. The strongest driving force is probably the environmental situation. As our impact on the environment starts to be both costly and a threat to our existence we shall have to look for new solutions for our energy supply. Solar radiation can be used completely without impact on the environment, and the flow of energy can be regarded as free. In addition the delivery is direct to the user, without any intermediaries.

We can already see today how some areas of use are competitive and as a result are expanding at a great rate.

FIGURE 11.7 *The largest solar cell plant in Sweden to date was installed in 1998 on IKEA in Älmhult. There are 380 m² of crystalline silicon solar cells mounted on the roof and 250 m² of thin-film solar cells (amorphous silicon) on the facade. The total power output of the plant is just under 50 kW, and it is estimated to produce 50,000 kWh per year*
Photo: Per Sandell, Neste Advanced Power Systems AB, Stockholm, Sweden

Both solar heating and solar electricity have an enormous potential for growth, and we can see that the market is expanding extremely rapidly at present. In a longer perspective solar energy will be a natural resource for our energy supply.

References and further reading

References

ASTIG (Active Thermal Industry Group) (2002) *Solar Thermal Markets in Europe*. Brussels, Belgium: ASTIG.

Claesson T *et al.* (1988) *Säsongslagrad solvärme i Kungälv*, Swedish Council for Building Research (BFR) report no. R104: 1988. Stockholm: Swedish Council for Building Research.

Dalenbäck J-O and Åsblad A (1994) *Förutsättningar för solvärme i gruppcentraler och mindre fjärrvärmesystem*, Proj nr 656–128–1. Stockholm: Swedish Business Development Agency (NUTEK).

IEA-PVPS (International Energy Agency Photovoltaic Power Systems Programme) (2002) *Trends in Photovoltaic Applications in Selected IEA Countries Between 1992 and 2001*, IEA-PVPS report T1–11: 2002. Swindon: Halcrow Gilbert.

Johnson R (ed.) (*1993*) *Solceller*. Stockholm: Swedish Business Development Agency (NUTEK).

Malbert B (1992) *Ekologiska utgångspunkter för planering och byggande*, Swedish Council for Building Research (BFR) report no. T35: 1992. Stockholm: Swedish Council for Building Research.

Perers B (1992) *Solvärme för badanläggningar: Projekteringshandledning*, Swedish Council for Building Research (BFR) report no T25: 1992. Stockholm: Swedish Council for Building Research.

Studsvik AB (1991) *Avdelningsrapport ED-91/5*. Nyköping, Affärsområde System och Distribution Studsvik AB.

Svenska Solenergiföreningen (1991) *Solsverige 1992*. Täby: Larsons Förlag.

Svenska Solenergiföreningen (1992) *Solsverige 1993*. Täby: Larsons Förlag.

Svärd G and Svård K (1992) *Klokboken*. Klippan: Pedagogförlaget.

Swedish Council for Building Research (BFR) (1994) *Termisk energilagring. Underlag för 1993–96*. Swedish Council for Building Research (BFR) report no G5: 1994. Stockholm: Swedish Council for Building Research.

Further reading

Behling, Sophia and Stefan (1996) *Solar Power: The Evolution of Solar Architecture*, Munich-New York: Prestel.

Butti K and Perlin J (1981) *A Golden Thread*. London: Boyars Pubs.

Dahm J. *Evaluation of a Solar Heating System for a Small Residential Building Area*. Chalmers University of Technology, Göteborg, Department of Building Services Engineering, Document D39:1997.

Dahm, J. *Small District Heating Systems*. Chalmers University of Technology, Göteborg, Department of Building Services Engineering, Document D48:1999.

Duffie J A and Beckman W A (1991) *Solar Engineering of Thermal Processes*, 2nd edn. New York: John Wiley & Sons.

Green M (1982) *Solar Cells: Operating Principles, Technology and System Application*, Englewood Cliffs, NJ: Prentice Hall.

Green, M (1995) *Silicon Solar Cells: Advanced principles and Practice*, Sydney, Australia: Australian Centre for Photovoltaic Devices and Systems.

Henning A (2000) *Ambiguous Artefacts: Solar Collectors in Swedish Contexts: On Processes of Cultural Modification*, Department of Social Anthropology, Stockholm University.

Herzog T (1996) *Solar Energy in Architecture and Urban Planning*, München: Prestel.

Johnson A M (1999) *Renewable Energy Technology: A New Swedish Growth Industry?* Chalmers University of Technology.

Markvart T (1981) *Solar Electricity*, Chichester: John Wiley & Sons.

Norton B (1992) *Solar Energy Thermal Technology*. Berlin: Springer-Verlag.

Rabl A (1985) *Active Solar Concentrators and their Applications*. Oxford University Press.

Wenham S et al (1995) *Applied Photovoltaics*, Sydney, Australia: Australian Centre for Photovoltaic Devices and Systems.

Zweibel K (1990) *Harnessing Solar Power: The Photovoltaics Challenge*. New York: Plenum.

For international information on solar energy, and for references to different countries' solar sections, I can warmly recommend contact with ISES, the International Solar Energy Society: www.ises.org

Information on accredited testing stations/ laboratories in Europe is presented at: ww.solarkeymark.org

Glossary

ΔT control (differential temperature controller)	Device able to detect a small temperature difference between different sensors and control pumps and other electrical devices accordingly.
absorber	Component of a solar collector that absorbs radiant energy and transfers this energy as heat into a fluid.
absorptance	Proportion of the radiation incident on a surface that is absorbed by that surface.
absorption	Process whereby a gas, liquid or a form of energy penetrates into and is taken up by a substance.
acrylic plastic	Collective term for rigid transparent amorphous plastics formed by the polymerization of polymethacrylate.
anodizing	Process in which a hard, non-corroding oxide film is deposited on aluminium or light alloys. The aluminium is made the anode in an electrolytic cell containing chromic or sulphuric acid.
auxiliary heat source	Source of heat, other than solar, used to supplement the output of a solar heating system.
combisystem	Combined solar DHW and space heating system using the same storage tank. Heat can also be supplied by another energy source.
concentrating solar collector	Collector that uses reflectors, lenses or other optical elements to redirect and concentrate the solar radiation passing through the aperture onto an absorber.
convection	Transfer of heat in a fluid (air/gas or liquid) by the circulation flow due to temperature differences.
convection barrier	Component in a solar collector that counteracts a cooling air movement between the absorber and the cover.
corrosion	Chemical reaction between a material and the surrounding environment. Often used to refer to the rusting of iron and iron alloys.
depreciation period	Calculated length of useful life of an investment.
DHW	Domestic hot water; the hot water that can be drawn off at a tap.
diffuse solar radiation	Solar radiation that reaches the ground via reflection and passage through the atmosphere and clouds; does not give strong shadows.
direct solar radiation	Solar radiation that reaches the ground directly; gives strong shadows.

direct system	Solar heating system in which the heated water that will ultimately be consumed by or circulated to the user passes directly through the collector.
double-jacketed tank	Double-walled tank in which the heat transfer fluid circulates in the space between the walls.
drainback system	Solar thermal system in which, as part of the normal working cycle, the heat transfer fluid is drained from the solar collector into a storage device when the pump is turned off and refills the collector when the pump is turned on again.
emittance	Measurement of heat radiated by a surface. (Also used to refer to light.)
EPDM rubber	Ethylene propylene terpolymer vulcanizable synthetic rubber.
evacuated tube collector	Collector employing transparent tubing (usually glass) with an evacuated space between the tube wall and the absorber.
finned-coil heat exchanger	Helical coil heat exchanger, normally of copper, in which the surface area is enlarged by fins.
flat-plate solar collector	Non-concentrating solar collector in which the absorbing surface is essentially planar.
galvanic corrosion	Corrosion resulting from the current flow between two dissimilar metals in contact with an electrolyte.
global radiation	Hemispherical solar radiation received by a horizontal plane
heat exchanger	Device specifically designed to transfer heat between two physically separated fluids. Heat exchangers can have either single or double walls.
heat load	Energy required at a given point in time, e.g. the heat needed during the coldest day (point in time) of the year.
heat transfer fluid	Fluid used to transfer thermal energy between components in a system.
indirect system	Solar heating system in which a heat transfer fluid other than the water ultimately consumed by or circulated to the user passes through the collector.
inhibitor	Substance that stops or reduces the speed of a chemical reaction. In a solar heating loop glycol mixed with water is used as the heat transfer fluid; an inhibitor is added to reduce the corrosive properties of the glycol.
insolation	Incident solar radiation; the solar radiation incident on a surface. The surface (plane) orientation must also be specified.
inverter	Device that converts direct current (d.c.) into alternating current (a.c.).
legionella bacteria	Bacterium, Legionella pneumophilia, which proliferates in stagnant water at temperatures under 50–60°C. Infection by droplet inhalation from air conditioners and showers can cause legionnaire's disease, which may be fatal.
peak power	Amount of power output by a photovoltaic module at standard test conditions (STC): module operating temperature of 25°C in full sunshine (irradiance) of 1000 W/m2. Unit: watt (W); also written Wp.
photovoltaic effect	Process that produces electricity directly from sunlight by the production of an electromagnetic field across the junction of dissimilar semiconductor materials when exposed to solar radiation.
power demand	Demand for a specific amount of electrical energy at a given point in time.
pressure drop	Power required to move a fluid a certain distance in an enclosed space. The pressure drop is affected by the properties of the fluid and the enclosure (pipe).

PVC	Polyvinyl chloride; the best known and most usual plastic used in construction work, for example in tubes, sheet and film.
reflector	Bright or white surface that throws back (reflects) incident solar radiation.
selective surface	Surface whose optical properties (reflectance, absorptance, transmittance and emittance) are wavelength-dependent. Surfaces with low emittance in the long-wave range and high absorptance in the short-wave range are frequently used in solar collector applications.
solar collector	Device designed to absorb solar radiation and transfer the thermal energy so produced to a fluid passing through it.
solar collector cover	Transparent (or translucent) material or materials that cover the absorber to reduce heat losses and provide weather protection.
solar collector efficiency	Ratio of the energy removed by the heat transfer fluid over a specified time period to the product of the defined collector area and the solar radiation incident on the collector for the same period under steady-state conditions.
solar collector loop	Circuit, including collectors, pump or fan, pipework and heat exchanger, used to transfer heat from the collectors to the heat storage device.
solar fraction	Energy supplied by the solar part of the system divided by the total system load.
sputtering	Process of applying a thin film in a vacuum chamber.
stagnation	Status of a collector or system when no heat is being removed by a heat transfer fluid.
stagnation temperature	Highest temperature that can occur in a solar collector when it is exposed to intensive solar radiation and not cooled.
storage tank	Tank for charging, storage and discharge of heat.
temperature guarantee	Source of energy (e.g. in the form of an electric heater) that guarantees a given minimum temperature.
temperature stratification	Distribution of temperature in a given volume, e.g. the vertical temperature distribution in a storage tank. As hot water has a lower density than cold water it is located in the upper part of a storage tank.
thermal length	Describes the characteristics of a heat exchanger with regard to size and the temperature difference on both sides.
thermal mass	Amount of heat energy in a given volume of a material between given temperature levels.
thermal solar energy	Conversion of solar radiation into heat, as opposed to solar electricity, which is generated by solar cells.
thermosiphon	Natural liquid circulation caused by small difference in density between a hot and a cold liquid.
thermosiphon system	System that utilises only density changes of the heat transfer fluid to achieve circulation between the collector and storage device or collector and heat exchanger. The heat store must be placed higher than the solar collector.
tow sealing	Method of sealing joints between pipes and components using tow (flax fibre) and a special paste.
vapour barrier	Layer that is not permeable by gas (water vapour).

Index

Printed and bound by CPI Group (UK) Ltd, Croydon, CR0 4YY

23/10/2024

01777678-0017